THE RED CANARY

Also by Tim Birkhead

The Cambridge Encyclopaedia of Ornithology (co-editor)

Sperm Competition in Birds

Great Auk Islands

The Magpies

Promiscuity

The Wisdom of Birds

Bird Sense

Ten Thousand Birds: Ornithology since Darwin
(with J. Wimpenny and R. Montgomerie)

THE RED CANARY

The Story of the First Genetically Engineered Animal

Tim Birkhead

B L O O M S B U R Y

NEW YORK • LONDON • NEW DELHI • SYDNEY

To Karl and Götz

Published by Bloomsbury USA, New York
Bloomsbury is a trademark of Bloomsbury Publishing Plc

All papers used by Bloomsbury USA are natural, recyclable products made from
wood grown in well-managed forests. The manufacturing processes conform
to the environmental regulations of the country of origin.

LIBRARY OF CONGRESS CATALOGING-IN-PUBLICATION DATA HAS BEEN APPLIED FOR.

ISBN: 978-1-62040-757-8

First published in Great Britain by Weidenfeld & Nicholson in 2003
This paperback edition published in 2014

1 3 5 7 9 10 8 6 4 2

Typeset by Hewer Text UK Ltd, Edinburgh
Printed and bound in the U.S.A. by Thomson-Shore Inc., Dexter, Michigan

Bloomsbury books may be purchased for business or promotional use. For information
on bulk purchases please contact Macmillan Corporate and Premium Sales Department
at specialmarkets@macmillan.com

Contents

Illustrations

Acknowledgements

I never imagined that writing about the red canary would take me so deeply into a world so different from the one I normally operate in. The history of bird keeping is not something that either public or university libraries have bothered about and I soon discovered that much of the information I needed was widely scattered and tricky to find, although I have to admit that the challenge of tracking down obscure material was an exhilarating one. No one has brought this eccentric material together before and, as well as widening my horizons, it also widened my circle of friends.

My main thanks are to two friends in particular, without whom this book would not have been possible: Karl Schulze-Hagen, a busy medic who lives in Mönchengladbach and whose other interests include ornithology, art and the history of science; and Götz Palfner, a research botanist in the Department of Animal and Plant Sciences at the University of Sheffield. Karl and Götz spent hours, days and weeks helping me, well beyond the call of duty, for which I am very grateful.

Despite what I have said about libraries, there were several who provided invaluable help. At the Alexander Library in the Zoology Department in Oxford, Linda Birch was, as always, extraordinarily helpful; I am also grateful to M. E. Butter; Peter Day at the Chatsworth Library; Ann Datta and Effie Warr at the British Museum Library in Tring and Ian Dawson at the Royal Society for the Protection of Birds library. Edith Sonnenschein provided useful information and images from the library at Vogelwarte Radolfzell, Germany. Elaine Dean of the inter-library loans department in the University of Sheffield deserves special thanks; as does Rosemarie Nief, at the Wiener Library, London, for tracking down copies of some of Hans Duncker's papers.

The personal libraries of various enthusiasts were vital sources of information and I am indebted to those who so generously allowed me to use their books, copied material or checked particular facts for me. These include Hans Classen, Jaques Faivre, Christian Lemee, Russell Liddiat, Rolf Nagels, Jochen Klähn and Joachim Seitz. I am especially grateful to Rolf Schlenker whose knowledge, enthusiasm and bookcases seemed limitless.

Tracking down details of Hans Duncker's fascinating life was a special challenge and I am grateful to the following for their assistance: A. Bäumer-Schleinkofer; Michael Birkmann; Eberhard Focke; Rolf Gramatzki who kindly provided a photograph of Duncker; Ernst Mayr who put me in contact with people who knew Hans Duncker, including Gerd von Wahlert, who turned out to be the only person I spoke to who had met Duncker and who was a wonderful and endlessly patient source of help. Joachim Seitz accompanied me to the Übersee-Museum in Bremen to go through the Duncker 'archive' and kindly translated material there for me. I also thank Professor H. Walter for his specialist knowledge of Duncker. At the Bremen Staatsarchiv I am enormously grateful to Monika Marschalck for her help in locating information on Duncker and his associates in Bremen.

The bird enthusiasts who allowed me into their world and answered my questions were a tremendous source of information and inspiration. In particular I thank the bird catchers Emiliano, Miguelito and Fernando who educated, entertained and fed me so splendidly in Spain. Elsewhere Mike Chambers, Arthur Jepson, Terry Roberts, Terry McCracken, Klaus Speicher and Sean Fitzpatrick in particular, all provided invaluable assistance. At *Cage and Aviary Birds* Donald Taylor and Ron Oxley especially were very helpful. I am also indebted to those bird keepers with a special interest in the history of their hobby including Wallace Dean, Huw Evans, John Scott, Jim Spring, Ray Steele and Adrien Taylor, who runs a canary website and who was a wonderful source of information; Geoff Walker, top colour canary man in Britain, provided hospitality and valuable information; I also thank David Whittaker, secretary of the UK Budgerigar Society, Kevin Wirick in the USA and Bob Yates for their help. Grant Watson deserves special thanks – president of the Colour Canary Breeders Association (CCBA) and the main red canary man in Sheffield, UK – he generously gave me the benefit of his knowledge and first-hand experience: Grant died while I was writing this book.

Other people who helped in various ways include: Antonio Arnaiz-Villena, Ishbel Avery, Gert Baeyens, Hans-Martin Berg, H. H. Bergmann, Marc Boccara, Martin Bossert, Bill Bourne, Michael de L. Brooke, Terry Burke, Michael Burleigh, Maxime Caillau, Alex Casa, Isabella Catadori, Clive Catchpole, Rosie Cornwallis who played the flageolet tunes for me from the books of Hervieux and Hammersely, Alistair Dawson, Françoise Dussor, Graham Elliot at the Royal Society for the Protection of Birds, Natalino Fenech who together with Joe Sultana helped me enormously in Malta, Anita Gamauf, Christoph Gasser, Armin Geus, Oliver Gilbert, Franz Goller, Ann Grant, Karen Green, Jose Gutierrez, Hansruedi Güttinger, Sabine Hackethal (who provided me with information about Lazarus Röting), Jürgen Haffer, Michael Hattaway, Alain Hennache, Peter Hill, Paul Hodges, Herbert Hoi who translated an entire Ph.D. thesis in one long sitting; Pete Hudson, Randal Keynes, Desmond King-Hele, Jo Kirby at the National Gallery who told me about pigments, David Hollingworth who provided invaluable help with the illustrations, Linda Kirk, Robert Knecht, Christoph Knogge who took me to St Andreasberg, Thomas Knogge, Walt Koenig, Maria Kroll who identified Madame la Princesse for me, Ole Larsen, Stefan Leitner, Kate Lessells who helped with translation and took me to the Castricum vinkenbaan, Robert Lindner, Spath Lothar, Peter Marler, R. Mathews, Richard A. May, Richard Mearns, Anders Pape Møller, Bill Mordue, Rolf Nagels, Marion Napoli, Paola Nini who translated French articles for me, Fernando Nottebohm, Jayne Pellatt, Pietro Pizzari, Tom Pizzari who helped in many different ways, Compton Reeves, Matt Ridley, Clarence Royston, Filip Santens who provided information and excellent hospitality in Flanders, H. Schwarzwalder, K. Sittman, Jon Slate, Peter Slater and Rienk Slings who made me welcome at 'his' vinkenbaan. Temmen publishers in Bremen were particularly helpful, Franciso Valera was extra ordinarily helpful, as were Wal van der Henk, Rinse Wassenaar and Roger Wilkinson.

Karl Schulze-Hagen, Götz Palfner, Tom Pizzari and Richard Wagner all read parts or all of this book in draft; I thank them for giving up their time and for their advice. Thanks also to my agent Felicity Bryan for her support and encouragement, and to my editors, Richard Milner at Weidenfeld & Nicolson and Bill Frucht at Basic Books for their perceptive and helpful suggestions. I am also grateful to Nick Humphrey and Bill Swainson at Bloomsbury for their help and encouragement with this new edition.

Finally, I thank my family, Miriam, Nick, Fran and Laurie, for their patience.

Preface to the 2014 Paperback Edition

More than ten years have passed since I started writing *The Red Canary*. That book was the culmination of something I had been thinking about for decades, but it also led to the development of something new. I have always been fascinated by birds, and have been lucky to be able to convert that passion into a scientific career. As a boy I went to Leeds Market on Saturday mornings to look at the rows of caged foreign birds, and occasionally bought some to keep in the aviary my father had built for me. An aviary was an excellent way of getting close to birds: it allowed me to watch them more intimately than I could ever achieve with wild birds. Their colourful plumage, extraordinary song and intriguing behaviour fired my juvenile imagination.

My interest in birds eventually became an obsession, and I suspect that my parents began to wonder whether their encouragement had misfired. The truth is that the hours that I spent studying both captive and wild birds paid off. That was my ornithological apprenticeship, though I was almost entirely self-taught. It has been suggested that to become expert requires 10,000 hours of practice. I spent many more hours than that watching birds as a boy, and it was this that eventually enabled me to get a job as a university lecturer and study birds for a living.

The Red Canary was an excuse to revisit my childhood and to explore the culture of bird keeping. In the 1960s the keeping of cage-birds was in decline, but still widespread – numerous birds were still advertised for sale in the weekly newspaper *Cage and Aviary Birds*. Much of my scientific research over the years has been conducted on the zebra finch, one of the most popular of cage birds. I have studied them in the wild in their native Australia, but also in captivity where they obligingly reproduce throughout the year. As a consequence I got to know many zebra finch breeders, some of whom were incredibly knowledgeable and helpful. They piqued my interest: what was it that generated their passion for bird breeding? How come these old men – for that's what almost all bird-keepers were – knew so much that I, as a professional ornithologist, didn't know? In some ways this was swampy ground, because bird-keeping is increasingly considered non-PC, and my scientific colleagues were often appalled at the idea that I would fraternise with those who they considered dubious company. Undeterred, I began to wonder whether ornithologists had missed a trick by avoiding and ostracising bird-keepers.

These ideas rattled around in my head for years, and almost subconsciously I began to search for a hook on which to arrange and develop them. Then, on a whim in the mid 1990s, I looked at a small book by A.K. Gill on canaries, written in the 1950s. Flipping through its pages in a second-hand book shop, I was shocked at Gill's literacy. Most books on bird keeping were poorly written, but this one was different. Gill was a librarian with a passion for canaries, including one I had never even heard of, a red canary. I was hooked.

The history of research on the red canary, described in the pages that follow, opened up a new world for me. Perhaps most importantly, it made me realise that the scientific study of birds began with bird-keepers in the mid 1600s. That led to a wider interest in how we know what we know about birds, and eventually resulted in the publication of *The Wisdom of Birds* (Bloomsbury, 2008), *Bird Sense* (Bloomsbury, 2011) and *Ten Thousand Birds: Ornithology since Darwin* (Princeton University Press, 2014).

Scientists get their inspiration from a wide variety of sources. Darwin, for example, maintained a voluminous correspondence with naturalists and animal breeders, joined two London pigeon clubs, and thought through his ideas by walking round the wood at the bottom of his garden. My first serious meeting with bird fanciers, described on page six, was part of my research for this book, and was both a cultural and scientific shock. A bird keeper told me something that led to a several years of research and a succession of scientific publications. The idea that I could be inspired by events well outside my normal day-to-day life was both enlightening and exciting. Imagine then, how I felt when I discovered that the red canary was the offspring of a collaboration between Hans Duncker – a scientist and Karl Reich – a bird-keeper, in the 1920s.

The Red Canary is the story of this extraordinary collaboration – a partnership that generated scientific discoveries and eventually a brand new bird – the red canary.

Tim Birkhead, December 2013

Preface

On the morning of 2 August 1921, a forty-year-old high-school teacher, Hans Duncker, set off from home through the prosperous and busy streets of the German town of Bremen. Stocky, moustached and bespectacled, Duncker was near the cathedral in the medieval centre of the city when he suddenly stopped in his tracks. Tilting his head slightly to one side, he closed his eyes and listened. People stared as they brushed past, but Duncker was transfixed. Somewhere out of sight a nightingale was singing. No nightingale ever sang in August and certainly not in the middle of town. Perplexed, Duncker made a mental note to come back and locate the enigmatic songster.

He never did, for the following week he met the owner of the mystery bird. Karl Reich, thirty-six years old, was a well-known bird keeper who in his twenties had been the first person in the world to make sound recordings of birdsongs. But it was no record that Duncker had heard; it was something even more extraordinary. The exquisite song came not from a nightingale but from a special canary that Reich had engineered through a decade of dedicated breeding.[1]

Duncker's meeting with Reich was so propitious and so successful that he felt compelled to write a detailed account of how it occurred. The canary had sung from the balcony of Reich's apartment. Seeing it and hearing Reich's tale of how he had created it, Hans Duncker was bewitched. He could not believe that Reich, a humble shop-keeper, could have produced such a bird. The scientific implications of Reich's ingenious canary-nightingale creation were enormous. Reich believed that he had somehow transferred to his canaries the heritable ability to sing a nightingale's song. But Duncker, a devout Darwinian, knew this couldn't be true and his own ingenious explanation drew on a subtle partnership of genes and environment that was years ahead of its time.

The nightingale-canary was just the starting point. The two men became firm friends and embarked on a roller-coaster journey of discovery. But it was not the songs of birds that became the main focus of their efforts, but their colour. With Reich's bird-keeping facilities and Duncker's scientific knowledge they conducted a full-blown study of the genetics of canary colours. Their results revealed not only that the canary's colours were controlled in a Mendelian manner, but that it must have taken previous generations of canary fanciers a century or more of selective breeding to transform the canary's wild ancestor from a dull green bird into the familiar yellow one. Fired by these findings, by great faith in the new field of genetics and subsequently by the generosity of a wealthy benefactor, Duncker devised the audacious and innovative plan to create a brand-new bird of his own: a red canary.

His idea was to pluck the genes from a red siskin – a relative of the canary from South America – insert them into an ordinary yellow canary and turn it red. Duncker half succeeded in this quest and half failed, for a reason that cuts to the very heart of the most important unresolved problem in modern biology. Duncker's genetic knowledge was unequalled and his logic impeccable, but despite years of trying he was never able to produce a truly red canary. He failed because his belief in genetics was so strong, so all-embracing and, ultimately, so naive. He came within a whisker of success: the canaries Duncker produced were a reddish coppery hue, but he never managed to breed birds of the intense crimson of the red siskin. The problem was that Duncker refused to recognise that colour could be determined by anything other than genes. There's a dreadful irony here, for when he previously figured out how Reich's canaries acquired their nightingale song, Duncker explicitly acknowledged the necessity of both an environmental and a genetic input. Had he applied the same reasoning to the birds' colour, he might have got what he was after. But when it came to feathers he had his heart set on a *genetically* red bird. It took others in more enlightened times to see beyond Duncker's genetic horizon.

Our understanding of the relation of genes, species and evolution (which is still, of course, far from complete) started in 1900 with the rediscovery of the ingenious experiments on inheritance in peas carried out during the 1860s by Gregor Mendel in the monastery gardens in Brno. Mendel's results, though published, lay unnoticed for almost forty years. Once recognised for what they were, they stimulated a tidal wave of new research as well as a torrent of controversy among investigators struggling to make sense of their results. It took two decades of backbiting and character assassination for researchers to sweep away the misunderstandings and to start moving forward on a united front. By the early 1920s the way was clear for a new and exciting phase of genetic endeavour.

Hans Duncker was to birds what Gregor Mendel was to peas. By combining his understanding of inheritance with the hidden knowledge of generations of bird keepers he was able to go one better than either Darwin or Mendel and start to build his own organism – a genetically engineered canary. He also precipitated a fervid worldwide contest among bird keepers to create a red canary.

A genetically engineered animal is one that has had one or more genes from another type of organism added to its genome by man. Today's technology is so sophisticated that in terms of putting the DNA from one organism inside another it doesn't matter how similar or dissimilar they are. Molecular biologists have no problem, for example, placing human genes inside a bacterium. But in the 1920s the only way to get the genes from one animal species inside another was by persuading two similar species to copulate, inseminate and fertilise each other. Even this wasn't sufficient: some of the resulting hybrid offspring had to be fertile. Today's red canary is a transgenic or genetically modified organism precisely because it contains one or more genes from the red siskin. Creating it involved crossing two species, repeatedly back-crossing their offspring to canaries so as to whittle away generation after generation of the unwanted siskin genes. The aim was to remove everything except the gene or genes that programmed feathers to turn red.

Compared with today's technology, hybridisation was an extraordinarily blunt instrument. But it worked and the proof in the form of genetically modified canaries exists today at every canary show. Still, the main reason Duncker failed to breed a red canary is easy to see today: genetic knowledge in the 1920s was simplistic. Mendel's rules were sound building blocks, but subsequent research has revealed an extraordinary number of demons – including imprinting, gene-environment interactions and maternal effects – that can distort Mendel's tidy ratios.[2] Attempting to create a red canary through hybridisation was rather like letting a proverbial monkey loose on a keyboard and expecting Shakespearean sonnets. But the red canary wasn't created by a mindless monkey. It emerged because by using the genetic knowledge then available together with a bit of luck the colour pioneers eventually found a canary genome that would allow the siskin's red genes to express themselves. But they needed more than genes to do it.

There was another factor in Duncker's failure: the *Zeitgeist*, which in Germany in the 1920s and 1930s was starting to turn horrifyingly ugly. The all-encompassing obsession with genes and genetic heritage that fired Duncker's quest to achieve a red canary by breeding alone also informed the political and social policies of the Third Reich: breeding, blood and purity of the genes determined everything from colour in a bird to courage in a man. Duncker's position as a high-school teacher coupled with his success as a geneticist meant that when the Nazis seized power in 1933, they were as keen to appropriate his expertise as he had been to appropriate the red siskin's genes a decade earlier. Inexorably, Hans Duncker was sucked into the ghastly Nazi machinery that proceeded to tear Europe apart over the succeeding years. Whether he did so willingly remains unclear.

The tale of the red canary is Duncker's story. But it embraces, too, the personal histories of a handful of others who were driven to transform a sombre little green bird into a flaming red one. This book is also an exploration of the curious interdependency of birds and humans over time and of what it is that drives men like Duncker to pursue a goal as esoteric as a red canary. It is a story that oscillates back and forth between the trivial – the quest for a red canary – and the deadly serious – the political abuse of biology – between good and evil, amateurs and professionals, love and hate and between past and present.

Hans Duncker's part in creating the red canary is a pivotal one, but his results and the success of his followers represent mere twigs on the ancient tree of bird-keeping tradition. Like most magnificent old trees, it continues to grow but it does so now only slowly. Bird keeping was at its most vigorous in the eighteenth and nineteenth centuries, and those periods of rapid growth and prolific flowering represent a remarkable and fascinating part of history when birds were kept with an almost obsessive zeal. Telling some of this story allows us to better appreciate Hans Duncker's achievements and to understand why the bird-keeping subculture is the shape it is.

Our relationship with genetics over the last hundred years has been horribly inconsistent. With the initial knowledge gleaned from the rediscovery of Mendel's work it was easy to elevate genetics to a position of supreme power and see it, as the eugenicists did, as the answer to all biological questions. Theirs was a Utopian vision and their goal was the improvement of the human race through better breeding. After the euphoric days of genetic optimism in the 1920s, events in Germany during the next two decades showed how spectacularly and dreadfully wrong the eugenicists had been. The study of genetics fell into disrepute. It took another thirty years for the subject of how 'genes' influence behaviour to re-emerge, albeit cautiously and in a succession of guises: including behavioural ecology in the 1970s, evolutionary psychology in the 1990s and genomics in the last few years. With the current obsession with genes – the human genome project, genetic engineering and transgenic wonders – we are tempted to regard the gene once again as a deity rather than a demon. In our optimistic awe of the power of genes we risk losing sight of the fact that without an environment to operate in genes are all but meaningless.

The Red Canary is the story of how, on their own, genes are not enough. It is a lesson in the history of biology. Understanding the intricate dialogue between nature and nurture – the interactions between genes and the environment – is the single most important biology lesson there is.

I

Igniting the Genome

There is a strong political coloration to the inclination to attribute inequalities of intellectual performance to nature rather than nurture ... The attempt to discover and promulgate the truth is nevertheless an obligation upon scientists ... for otherwise what is the point of being a scientist? The alternative is for scientists to content themselves with being the mere handymen and artisans of a machine-based culture.

P. B. MEDAWAR and J. S. MEDAWAR, *Aristotle to Zoos* (1984)

Inside a tiny cage a small, blood-red bird flutters and calls. It is a *pitadour*: a decoy, a prisoner whose frantic attempts to escape merely draw others to the same fate. Around the decoy's cage are a dozen sticks smeared with bird lime – the most vicious tarbaby glue you can imagine. A wild bird, lured by the decoy's cries, comes close and alights on a limed stick. As it does so it falls to the ground, almost as though it has been shot. Lying helplessly in the grass, its feet and wings glued firmly to the stick, the bird is a picture of utter bewilderment. The trapper runs up and collects his prize, pulling away the limed twig and spitting on to the bird where patches of lime remain. Using his thumb and index finger he rolls the lime off the bird's plumage into tiny balls, flicking them away as he walks back to his blind. The bird is a red siskin *Carduelis cucullata* and known locally as *cardenalito* – little cardinal – a tiny, vivid vermilion finch with a black face. This one is a male and a good one, its plumage the deep scarlet of oxygenated blood, and worth much more than the paler birds the man trapped earlier. He thrusts the bird into a low, flat wire box – a keeping cage – which prevents its occupants from jumping up and damaging themselves. Inside, the captives huddle together in the corner – all facing outwards with their feathers flattened, and all gasping with fear.

The man returns to his hide and almost immediately two female siskins alight on the limed twigs. He runs forward again but is irritated. Females are worthless – their plumage is grey and anaemic compared with the males' – and dealers don't want them. Carelessly he tears the two birds from the limed twigs. He releases them but deliberately pulls out their tail feathers as he does so and throws the birds into the vegetation, where they scuttle off like mice. The trapper believes that the females, unable to fly, will help to pull other males into the area, but in reality the flightless females are doomed and will soon be killed and eaten by predators.

The northern part of Venezuela is a region of rolling hills and scattered shrubs, with small *cafetals* – coffee plantations – gardens and areas for stock grazing. In the distance lies the blue haze of the Caribbean and on it the faint outline of Curaçao. It is 1925 and the trapper cannot believe his luck – everything is going so well for him. His father, and his father before him, trapped siskins and other birds here, and sold them to dealers who shipped them off to the Canary Islands, a Spanish colony, 3000 miles away. In his father's day trapping barely paid for itself, but it was fun outwitting the birds. Today, as our man lifts the holding cage and shields his eyes against the sun to look at the morning's catch, bird trapping is positively lucrative. This isn't because the price of individual birds has risen but because the market now seems limitless: dealers simply cannot get enough birds and our man is an extremely skilled trapper. Years of watching and working alongside his father have paid off. He knows exactly where to place his limed twigs and precisely how they should be positioned to catch the birds without damaging them. Other trappers in the village, keen to capitalise on the booming bird market, are less adept and they capture many fewer birds, often damaging those they do catch. Our trapper's success has made him something of a local celebrity; his trapping achievements have put money in his pockets and several señoritas have given him the eye in the hope that they might eventually get their hands in his pockets.[1]

Later in the day the birds are collected by a dealer who drives them down to the coast. There they join dozens of other increasingly stressed individuals crammed together into cages without food or water. Towards dusk, as the temperature falls, a man comes by and pours a bucketful of seed through the tops of the cages, and attaches a small drinking pot to the wire front of each cage. As he walks away the birds crowd forward to suck desperately at the water. But already some of the birds are dead, and lie rusty red among the filth and feathers on the floors of cages. After dark the birds are placed in the hold of a ship, along with cages of parrots, finches and a variety of monkeys, to begin their journey to the Canary Islands and for some of them, ultimately, to Germany.

Six weeks later in the leafy suburbs of Bremen, one of Germany's two main ports, Hans Duncker, high-school teacher and canary enthusiast, eagerly awaited the six red siskins he had ordered. Like many of his countrymen Duncker was passionate about birds. He was interested in everything and anything to do with them, and having written a prize-winning book on birds twenty years earlier, he has a well-deserved reputation as a scientific ornithologist. Duncker's life was changed after World War I by his chance meeting with champion canary breeder Karl Reich. Together they hatched a plan to create a red canary by appropriating the genes from a red siskin and putting them inside an ordinary canary.

Reich telephoned to tell Duncker that the ship bearing the siskins had arrived and he set off for the port to collect them. Of the original 700 birds which left Venezuela over 600 had died en route, most of them within a few days of capture. The Bremen bird dealer, keen to dispose of the survivors as quickly as possible, caught six individuals with his bare hands and placed them in the small cage Reich had taken with him. Starting for home, past the coffee and tea importers' elegant warehouses lining the east bank of the Weser river, Reich was eager to release the birds into their new home. They weren't cheap, but if their plans came to fruition, these little red birds would more than pay for themselves. Reich arrived at his home on the edge of the old town where Duncker was waiting for him and together they went into his bird room to release the captives into a big cage. Only five siskins flew out. The sixth lay dead, like a blood smear on the bottom of the box. The trauma of the dealer's hands had been the final insult. Reich mentally shrugged; like all bird keepers he knew that mortality was high among new acquisitions; it was survival of the fittest and the sight of dead birds was something you got used to.

Startled by the new sense of space, the survivors flew frantically back and forth inside the cage. After a few moments they stopped and, breathing hard, clung to wire to look nervously at their new owners. Reich and Duncker peered back at them. All the birds were males, but four of them had dull red plumage with badly soiled feathers. But the fifth bird, against all the odds, was immaculate. Somehow, through weeks of travel in cramped and crappy quarters, this one individual had managed to keep itself in perfect plumage. This is what Duncker and Reich had waited for. Their genetic experiments could now begin in earnest.

The Venezuelan trapper who removed the red siskins from their native forests for Duncker and Reich was one of thousands across the globe who caught birds for the booming European markets. Across Africa, India and South America bird catchers supplied an insatiable appetite for small birds, fuelling a forced migration of feathered bodies into European cities. The practice of keeping small birds in cages was an ancient one, dating back almost to the beginning of civilisation, reaching a peak during the late 1800s.[2] By the 1920s, when the possibility of creating a red canary first ignited Duncker's imagination, cage birds were still popular and virtually ubiquitous across much of Europe.

Germany was the stronghold of European bird keeping and had been so for centuries. It was here, 400 years earlier, that ingenious and dedicated bird enthusiasts had first started to create a yellow canary, and the German tradition of genetic tinkering had continued ever since. Attitudes towards keeping and breeding birds, however, had changed and nowhere more so than in Britain. The British were still very keen on cage birds in the 1920s, but the hobby was less popular than it had once been. People still kept canaries, parrots, foreign finches and more recently grass parakeets, also known as budgerigars, but the habit of keeping birds, and native species in particular, was in retreat. Ever since the end of the nineteenth century when a small group of wealthy women started to question the ethics of keeping little birds – especially those from their own country – the catching and caging of birds had been under threat. These women campaigners were typical bourgeois products of the colonial age: less concerned with the welfare of foreign birds, they were adamant that keeping native species in cages was wholly inappropriate. Thus started in England what would later become the Royal Society for the Protection of Birds (RSPB) and the slow but sure decline in British bird keeping. So successful was the RSPB's anti-bird-keeping campaign that today bird keeping is a mere shadow of its former self and, in all but a few sectors of society, viewed with great suspicion. It is worse than that. In Britain keeping native species is now considered a disreputable hobby and admitting to being a bird keeper is like confessing a crime. In Germany things could not be more different. Keeping birds is still a respectable pastime and aviculture has retained its respectability. Professional bird researchers there recognised a long time ago, and continue to recognise, that the accumulated knowledge of generations of bird keepers might actually be useful, not just for ornithologists but for science in general.

Science and amateur bird keeping are parallel cultures and, like lines that are almost parallel, they rarely meet, but when they do, as Hans Duncker later discovered, extraordinary things can happen. My own quest for the red canary took me across Europe, from the cradle of canary culture in Germany's deep south, to Bremen in the north, where Duncker spent his working life. I oscillated between the poorly lit basements of university libraries and the beautiful homes of private collectors who lust over rare bird books, to some seedy pubs and the more modest homes of bird keepers themselves. Entering the world of the fanatical bird keeper – those who enjoy mating different species together and those who bask in the reflected glory of their beautiful red canaries – made me feel a bit like Charles Darwin, who in the 1850s joined various pigeon clubs and sat among ardent fanciers in London's gin palaces in order to see what it was all about. We know only a little about Darwin's meetings with pigeon-fanciers, but I hope his first encounter was better than mine.

Soon after deciding to write this book, I was asked if I would talk to a bird breeders club in northern England. Their meetings, it turned out, were held in a public house in one of Yorkshire's less attractive cities. It was dark and raining as I arrived on the appointed evening, acutely aware that this was an area where I wouldn't normally consider getting out of my car, let alone going inside a pub. But I did, and carrying my projector and screen I felt a bit like a conjuror turning up at a children's party. My reception was distinctly frosty and I began to wonder why I had bothered. Standing beside the projector, I waited to be introduced, much as I would at a scientific meeting, when, with more than a touch of impatience, someone shouted from across the room, 'Well, go on, then!'

The body language of the audience was hardly encouraging: to a man they were leaning back in their chairs, as far away from me as they could get. Years of recalcitrant undergraduate audiences were nothing compared with this, but I carried on and told them about my research – how sperm meets egg and how they might breed better birds. And slowly I sensed their resistance melting away. Some of them even started to lean forward in their seats. After forty minutes there was a break and they all trooped downstairs to refill their glasses; meanwhile the barman appeared with an enormous plate of sand-wiches and pork pies. 'Tuck in, lad,' someone said, so I did and I noticed that someone had bought me a pint. Fifteen minutes later, after everyone had refuelled and peed, we were off again. Somehow the atmosphere seemed different. By the end of the talk there was no question: they had enjoyed it and the discussion continued unabated for nearly an hour. I felt almost euphoric. The contrast with the situation a few hours earlier was unbelievable. And just as I was about to leave, someone asked me if I would give the same talk to the lads at *his* club. A few weeks later I did just that and, to my utter amazement, almost everyone who had been at the first meeting was there as well as many new faces. Hans Duncker – dedicated to bridging the gap between professional biologists and bird keepers – would have been proud of me.

My distinctly chilly reception at this, my first, bird keepers' meet-ing was, I later realised, something I should have anticipated. British bird keepers live in fear of the law and are suspicious of anyone from the outside. The bird protection bodies have done an outstanding job in persuading people – mainly middle-class people – that keeping birds, mainly by working-class people, is socially unacceptable.[3] In Darwin's day, keeping pigeons or canaries was both respectable and classless. And the same was true in Duncker's Germany.

Duncker knew it and Darwin knew it, but few others before or since have appreciated the scientific potential of the vast store of untapped knowledge hidden away in the collective memory of bird keepers. The long history of bird keeping, stretching back in time, was the crucial strand of knowledge on which Duncker's and Reich's success would ultimately rest.

A bird keeper since boyhood, Reich was the repository of that knowledge. Through his writing, recordings and public performances he had already passed on some of that information to a new generation of bird keepers. While this helped to maintain the avicultural tradition, on its own it was rather like cloning or asexual reproduction: it simply generated more of the same but nothing novel. The combination of Reich's knowledge and Duncker's imagination, on the other hand, was loaded with potential from the very beginning. Reich was the cultural transmitter of bird-keeping know-how, enabling Hans Duncker to meld it with his own specialist understanding of heredity. Their respective sets of knowledge were like chromosomes from different parents coming together during fertilisation, crossing over and recombining to provide a combination of talent novel enough to change the world.

2

Catching, Keeping and Status

I shall use my endeavour to give you an account of all the hard-beaked Bird wich feed upon Seeds, and are most plentiful with us here in England; the first I shall begin withal is the Bird called the Canary-Bird, because the Original of that bird came from thence (I hold this to be the best Song-Bird); But now with industry they breed them very plentifully in Germany, and in Italy also; and they have bred some few here in England, though as yet not anything to the purpose as they do in other Countries.

JOSEPH BLAGRAVE, *Epitome of the Art of Husbandry* (1675)

Hans Duncker was born on 26 May 1881, some three years before Gregor Mendel died. Duncker's birthplace was the town of Ballenstedt in the eastern foothills of the Harz Mountains in northern Germany, the very core of German bird-keeping country.[1] People here had caught and kept birds, and exploited the singing abilities of species like the chaffinch, since the Middle Ages. When the canary first came to the region in the late 1700s, miners and artisans put their hearts and minds into making it the ultimate singing bird – one that eventually became known worldwide as the Harz Mountain roller. The scale on which they eventually did this is almost inconceivable. By the 1880s, in one of the most remarkable cottage industries in history, Harz families were fledging more than 150,000 male canaries each year.

The second of three sons, Hans Duncker grew up amidst this bird-keeping culture. Everywhere he went there were caged birds, a colourful, vociferous array of wild-caught birds and roller canaries; in doorways, hanging outside houses, in shops and in cafés there were singing birds. His grandfather encouraged his interest in birds and took Hans to family friends who had aviaries full of native finches and canaries – wonderful combinations of colour and voice. There were siskins and serins, wild canaries and greenfinches whose plumage was as bright as fresh foliage in dappled sunlight; there were chaffinches, redpolls, linnets and bullfinches whose breasts were as pink as the blushes of teenage girls; and then there was the favourite, the multicoloured goldfinch – a blaze of crimson and gold, black, white and beige – twisting and turning almost like a clockwork toy. Even the canaries' brilliant yellow plumage didn't seem out of place. But it wasn't just their appearance that was inspiring, the continuous birdsong made everyone in Ballenstedt feel as though they were living in the forests and fields. The music of nature was spiritually uplifting and almost everyone was an enthusiast. Men returning home from work in the evening went to their birds before going into the house to greet wife and family. It was hardly surprising, then, that young Hans, shy and somewhat serious, should turn to birds for his studies. His father, a high-ranking judge, decided that if Hans wanted to study birds he should do it properly and train as a biologist. It was a long slog: birds, and cage birds in particular, were the domain of amateurs rather than professional scientists. In order to get to a position where he could focus his efforts entirely on them, Hans had first to serve an apprenticeship. He did well at school, especially in maths, physics, sports and singing, but he hated languages and found French and English a struggle. At eighteen he was sent off to the other side of the mountains to study biology. The university at Göttingen was not only the closest but among the oldest and the very best in Germany. Its medieval half-timbered buildings dominated the small town, as did its elite fraternities of male students who sought recognition not only by their success in examinations but also by the sword, their facial scars the ultimate signals of status.[2]

Initially Duncker concentrated on botany and physics, but after a year-long visit to Leipzig University in 1901 he was able to begin his zoological studies. Returning to Göttingen in the Easter of 1902 he sought out Ernst Ehlers, grandseigneur of animal structure and close friend of Ernst Haeckel, Germany's greatest zoologist and most ardent Darwinian. One of Haeckel's many claims to fame was his idea of plotting the branching paths of common descent on to evolutionary trees, and he entreated Ehlers and his students to do the same. Ehlers in turn encouraged students like Duncker to read Haeckel's monumental works, including his popular versions of Darwin's ideas. By examining the internal structure of different animals, mainly invertebrates such as slugs, snails and worms, Ehlers could judge how closely related they were and then position them on the appropriate branches of the tree of life. Zoology in the early 1900s comprised little more than the detailed study of morphology – and deciding whether animals had been put together by God or natural selection. For Ehlers, as for Haeckel, there was no question: natural selection had created life's wonders. Ehlers gave Duncker a doctoral project that involved identifying the relationship between a few more twigs on the tree of life. The work entailed comparing, dissecting and describing the internal structure of a group of marine worms. This wasn't quite as dull as it first sounds, for these were among the most gorgeous of all worms – affectionately known as 'sea mice' for their squat and hairy appearance. Linnaeus had been so impressed by their iridescent beauty that he christened the commonest form 'Aphrodite'. Ehlers used the worm project to drill into Hans the scientists' 'right stuff': persistence, integrity and imagination.[3]

Away from Göttingen's laboratories and lecture theatres, Duncker was an active member of 'Germania', the National Christian student fraternity, and another Christian society, 'Schwartzburgbund', which approved of chastity and disapproved of duelling. Despite his

Christian upbringing, Duncker was converted to Darwinism by Ehlers and remained a firm believer in evolution by natural selection for the rest of his life. He was a model student and did well.[4] Following his oral examination on 15 February 1905, he graduated *magna cum laude* – the second-best grade – in zoology, botany and mathematics. Thereafter he was Herr Doktor Duncker.

On their own, worms were not enough, and between dissections Hans continued to pursue his interest in birds – in particular, their migrations. For centuries it had been known that entire populations of birds appeared and disappeared at predictable times of year, but only in the past few years had there been any attempt to assess these movements in a scientific way. The first bird observatory, at Rossitten in East Prussia, had been set up in 1903 specifically to employ the new technique of marking birds with metal rings to track their subsequent movements. The establishment of this ringing station helped to make bird migration a hot topic once again, even though Duncker was personally sceptical about the value of ringing birds, Nonetheless, in the year before he completed his thesis Friedrich Voss, a friend, encouraged Duncker to give a talk on migration. Always conscientious, Hans hurried off to the university's magnificent, ancient library to swat up on the subject and nervously prepare for his first public performance. He need not have worried. He had a natural flair for public speaking and with the migratory journeys of birds one of the most extraordinary aspects of biology, his talk was an enormous success. Encouraged by Ehlers, Duncker decided to turn his lecture into something more substantial, and was soon back in the library searching for new information and plotting the migration paths of birds across the entire northern hemisphere.

In the Blood

For centuries men have gouged silver, iron and coal from beneath the Harz Mountains. The same men have also kept birds. Birds were the miners' saviours and for centuries locally caught chaffinches had been carried underground to warn of poisonous gas. Once canaries arrived in the Harz region around 1800 they took over this role. A canary in a coal mine became an enduring image: a speck of life in dreadful, dangerous blackness, whose high metabolic rate and sophisticated respiratory system made it supremely sensitive to poison gas. The merest whiff of carbon monoxide or fire-damp – an explosive mixture of air and methane – would knock a canary fluttering from its perch, providing its owner with a clear signal to retreat. The long history of birds underground forged an almost unbreakable chain of mutual dependence between miners and songbirds.[5]

German bird keeping extended well beyond the Harz Mountains and well beyond those who toiled underground. By the late 1800s keeping small birds in cages was a national obsession. In medieval times the catching of small birds was one of four noble and status-enhancing pursuits along with hunting, falconry and fishing.[6] A hierarchy even existed among these different activities, with men who killed deer or boars deemed braver than those who used falcons and hawks to take partridges, who in turn were more worthy than those who plucked little birds from the sky. Fishing, the most solitary and static of these activities, conferred the least status. Nonetheless, all who partook in these sports regarded themselves superior to common peasants.

Books describing techniques for each form of hunting proliferated across Europe, especially in England. In his little book, *Hungers Prevention*, published in 1621, Gervase Markham[7] said that small birds have two uses: 'either pleasure or food, pleasure because everyone of them naturally, have excellent Fielde-Notes, and may therefore be kept in cages and nourisht in their owne tunes, or also trayned to any other notes, or else for food, being of pleasant taste, and exceeding much nourishing, by reason of their Naturelle heat, and light digestion'.

Most small birds fell into one or the other of Markham's categories: for eating or keeping, but larks had the misfortune to have both a wonderful song and a wonderful flavour – their tongues (in reality their breasts) were considered the most succulent of all meat – and larks became a prime target for bird catchers everywhere. The Italian historian and cleric Polydore Vergil,[8] who lived in England between 1501 and 1550, wrote, 'The cheefe foode of the Englisheman consisteth in fleshe . . . Of wilde burdes these are most delicate, partriches, pheasunts, quayles, owsels, thrusshes and larckes. This last burde, in winter season, the wether being not owtragios, doth waxe wonerus fatte, at which time a wonderful nombre of them is caughte, so that of all others they chefle garnish menns tables.'

Although eating was an important end product of catching, a common motivation, as in any form of hunting, was success in catching and the status it conferred. The quest for status inserts itself into all pursuits; catching small birds was no different and the result was a rapid evolution of catching techniques. Like species themselves, these evolved from a few simple forms until, by the Middle Ages, bird trappers employed a staggering array of bizarre and dreadfully effective techniques. Methods that failed became extinct, while those that succeeded, like their owners, proliferated.

The best time to catch birds was when they migrated, because they usually did so in huge numbers and in predictable places. As birds move from their summer breeding grounds to wintering locations and back again they funnel in vast numbers through mountain passes, along headlands and across islands. Autumn is the best time of all for catching because as the birds move south for some reason they do so on a much narrower front than they do on their return journey. October and November were the peak trapping periods, and if you look hard at Pieter Bruegel's painting *The Return of the Herd*, part of a series known as 'The Months', you can see – near its centre – a bird catcher concealed behind a bush, beside his net. The bird catcher is one of Bruegel's time markers. The other marker, just in case there was any doubt, is the grape harvest in the bottom right-hand corner, for this was October. There was an added bonus in hunting migrants. As they travel from their winter to summer quarters and back again, small birds are loaded with fat – fuel for the journey – and some more than double their weight with fat before setting off, making them especially tasty.

Throughout October and November, thrushes and larks move from northern Europe into Spain and France, and finches and warblers swoop off the toe of Italy, over Malta and on to North Africa. Millions of birds, large and small, migrate each autumn over the Mediterranean, crossing at narrows and using islands as stepping stones to minimise the risks. Every Mediterranean island was a hotbed of hunters. Some, like Malta and Cyprus, still are.[9]

The gourmand's favourite small bird was, and still is, the ortolan bunting;[10] and in the past vast numbers of these birds were captured alive and fed oats and millet seed until they became 'lumps of fat three ounces in weight'. Now they are caught – illegally – for specialist French restaurants to satisfy clients who each year eagerly await the discreet telephone call to let them know the birds are in. Behind locked doors they assemble for this sensual gastronomic ritual. A woman who had been to one of these gatherings described to me the technique for eating a cooked bunting: 'I was told very firmly to put the whole very hot bird in my mouth and then to press it up against the roof of my mouth with my tongue, rather than biting it. This makes the juice and the fat explode into your mouth, which I have to say is the best bit of the whole experience, since all you are left with after that is a mouthful of fine bones.'

Vogelherds

Duncker's lecture and his subsequent book on bird migration were based on knowledge that came originally from bird catchers. For these men knowing about migration was no mere intellectual exercise but a matter of life and death. To be successful, bird catchers needed to know where and when birds would be moving. Natural selection operated on trapping sites just as much as it did on trapping methods and the trappers themselves. The Italians' gustatory obsession with birds drove them to create leafy replicas of Stonehenge, known as *roccoli* – gigantic bird-catching structures – dotted across the mountain passes of northern Italy, whose locations tracked with deadly precision the exact migration paths of small birds as they crossed the Alps. In Germany and the Low Countries trappers used a different method – huge clap nets – established at fixed locations known as *Vogelherds* – bird yards or, tellingly, bird ovens.[11]

The vital role of bird catching in local economies is reflected in the huge number of places across Europe whose names are linked with trapping.[12] In Germany literally hundreds of places have names like *Finken-feld* (finch field), *Lerchen-feld* (lark field), *Lerchen-berg* (lark mountain) and *Vogelherd* (bird oven). In Britain such names occur less frequently because, unlike their German counterparts who maintained traditional fixed trapping sites, the British bird catcher was more mobile and opportunistic. Nonetheless, English bird trapping is immortalised by such places as Finchingfield and Larkhill in Essex.

The catching and keeping of birds, and observing them in captivity, were the beginnings of ornithology. Like all successful hunters, bird catchers had to know their quarry; they had to know first and foremost how to distinguish the different species – something we take for granted today, but before the Late Middle Ages often a puzzle – especially in those species where males and females differ in their plumage. Trappers had to know where wild birds would be at particular times of the year, what they ate, and how they responded to members of their own and other species. Such knowledge was the beginning of avian ecology. The careful observations of caged birds – in particular how and when they sang – provided the start of the study of animal behaviour, and observations of where birds were and went, the inception of the study of migration.[13]

Bird catching is illegal now across much of Europe, and those *roccoli* and *Vogelherds* that survive have become respectable bird-ringing stations for the ongoing study of migration. The old catching methods have all but disappeared, replaced by the amazingly effective Japanese mist nets, but the motivation is the same and even though few bird ringers would admit it, the real buzz is still the catching. In a few places, such as Castricum on the Dutch coast, the old *Vogelherd* clap nets are still used to catch birds in a kind of living museum. When I arrived one mild October morning before dawn I found the bird catchers already at work – their nets were set and they were waiting for the flocks of migrants that would set off down the coast as soon as there was sufficient light. As the dawn broke I could see the huge nets lying flat on the ground about twenty metres from the little hut in which we stood. Surrounding the nets were small cages, each one containing a different bird. In the past these decoy birds would have been blinded – making them easier to manage and less easily distracted from their main purpose of singing.[14] The effect of the decoys was almost uncanny. Waves of migrants appeared from the north flying low across the dunes, but the singing decoys caused them to hesitate and swoop down for a closer look. As they did so the 'catcher', who controlled the cable running out to the net, had to decide when to pull. If he got it wrong, the birds veered safely off and precious catching time was wasted resetting the nets. If he got it right, the two halves of the net swept over like a pair of gigantic jaws, plucking the birds from the air and pinning them to the ground.

We watched through the tiny windows as a flock of small birds approached. The decoys called; beguiled, the migrants descended and momentarily fluttered over the nets. The man pulled. Success! We ran out and picked up the pipits struggling beneath the nets and handed them to the ringers. Ours was a modest catch – just three birds. But in the past these trapping sites, known as '*vinkenbaan*' – the Dutch equivalent of *Vogelherd* – were sometimes much more successful. *Vinkenbaans* were the property of wealthy landowners who employed bird catchers to provide birds for the table. The birds were considered a crop and catchers were required to keep meticulous records of everything they caught. For one trapper who worked on the estate of Cornelius van Lennep, 9 October 1790 was a particularly memorable day: in a single pull of the net he caught 203 birds.[15]

Birds, Sex and Status

The catching and keeping of birds has always been associated with status and, among males at least, status has always been linked with sex. The acquisition and ownership of birds provide wonderful examples of the power of sexual selection to drive men to extreme behaviour. As with other hunters, the more successful a bird catcher, the greater his status in the local community, and men also gave birds to women as courtship gifts.

Agnès Sorel was a medieval beauty, with a face and figure to die for. She was smart, too. Joining the household of Charles VII, King of France, as a teenage servant in 1443, Agnès used her looks to catch the royal eye. I can only imagine that for the king it must have been like having Nigella Lawson in the kitchen and inevitably he became besotted by her. Other members of the court were extremely critical of Agnès's tactics, referring to her as the 'begetter and inventor of all that can lead to *ribaudise* and dissoluteness in the manner of costume', by which they meant she wore extremely revealing dresses, which exposed her ample bosom. Charles wooed her, by giving her a canary – along with several castles and estates – and Agnès soon became the first royal mistress. Canaries were rare, having just appeared in Europe and they were the ultimate status symbol, for both giver and receiver. The canary was also a deeply symbolic gift. Men gave them and women adored them – enchanted by their sweet voice and endearing nature. Agnès loved the canary and Charles too, bearing him four children and eventually becoming more powerful than the queen. For years after her suspiciously premature death in 1450, her beauty was celebrated in French paintings of the Madonna, usually suckling the infant Christ, providing a perfect excuse for displaying her magnificent figure in what now seem to be curiously ambiguous religious images.[16]

Agnès Sorel and Charles VII provide a revealing object lesson in contemporary evolutionary psychology. Agnès was everything the forty-year-old king could ever wish for; she was young, beautiful and irresistibly sexy. For his part Charles was everything Agnès could ever want: immensely rich and very powerful. Evolutionary psychologists tell us that the obsession of men with young and beautiful women is entirely explicable in Darwinian terms. A woman's good looks usually go hand in hand with youth and youth denotes fertility. Men prefer younger women because they'll give them the best return in terms of offspring and beautiful women make healthy, fertile partners. For a woman, a man's looks are almost immaterial – just think of all those ugly old men with beautiful wives or mistresses. What women want in a man is money and power, two attributes that usually go together. Men with resources determine a woman's reproductive output. With access to plenty of money, women can produce lots of healthy babies and ensure that they have the best possible upbringing, education and care. And it is precisely for this reason that the old men with beautiful partners are always rich.

While women strive to be beautiful and – literally – attractive, men aspire to be powerful.[17] Men need status and prestige to gain power. Not every man can be king and not every woman can be as beautiful as Agnès Sorel, but each of us does the best we can in the mate acquisition game – either consciously or unconsciously. For a man, getting a partner is all about competing with other men for status and then displaying that status in some way. Acquiring a partner can also be competitive for women, but it is more about choosing between different men. This difference in the mate acquisition strategies of each sex is Darwin's idea of sexual selection and he was well aware of the central importance of male status for both animals and humans alike: 'Man is the rival of other men; he delights in competition, and this leads to ambition which passes too easily into selfishness. These latter qualities seem to be his natural and unfortunate birthright.'[18]

His views were undoubtedly shaped by Victorian sexism and Darwin considered the creative achievements of men as equivalent to a peacock's tail or, more precisely, the bower bird's bower – an extension of themselves – and what Richard Dawkins calls an extended phenotype.[19] Catching birds was a male display, but so was owning them, and a bird that was both attractive and unusual – like an albino, a rare hybrid or a red canary – was a wonderful extended phenotype. Bird catchers enjoyed a special kind of prestige allowing them to reap its most important biological reward. In Germany and the Netherlands, and probably elsewhere, bird catchers had a reputation as womanisers, not least because they often took young women with them to their remote trapping locations. This may explain a curious imagery that arose in sixteenth- and seventeenth-century art, when bird trapping became a symbol for love and sexual intercourse. A caged bird – in Dutch 'vogel' – symbolised the happy slavery of someone in love, whereas, 'vogelen' was the vernacular term for the sexual act. This connotation of bird trapping appears in numerous European works of art and is particularly explicit in Jan Steen's painting *Rustic Love*, completed around 1660. The picture shows a young couple frolicking in the open air; an enthusiastic young man about to capture a young woman. Above them in a tree is a cage trap, with its trapdoor wide open and decoy bird inside.[20] The same kind of double entendre occurs many times over in Mozart's opera *The Magic Flute* in which the part of Papageno, the bird catcher, is another play on the German 'Vogel' and 'vögeln', just as it is in Dutch. Papageno's song at the start of the opera goes:

> The bird catcher am I and always merry, tra la la!
> As the birdcatcher I am known, by young and old throughout the land.
> I know how to set decoys and whistle just like my prey.
> So merry and carefree can I be, knowing all the birds belong to me.

All the birds belong to me. Even today birdwatchers in Germany are confronted with this sexual innuendo every time they tell others about their hobby. *Vogel* means bird and *vögeln* means to fuck so 'going birding' in Germany is a dubious pursuit. The same thing exists in Britain, too, where a 'bird' or a 'chick' refers to a girl – a dual meaning that also dates back to the heyday of bird catching – even the words 'bird' and 'bride' have the same root.[21] Pulling a bird – an English expression denoting a man's success at initiating a liaison with a woman – refers to the act of pulling the clap net ropes to capture birds, just as the phrase 'picking up a bird' refers to what one did after a bird was caught under the clap net. In Italy, the sexual innuendo is even more explicit. The word for bird, *uccello* is slang for 'penis', much like 'cock' in English, and *passera*, literally 'little sparrow', refers to the vagina.

The link between bird catching and sex persists to this day. Once, when I was in a remote and mountainous part of Spain with two old bird trappers, one of the men caught a bright-red cock linnet. After disentangling it from the sticky bird lime with which it had been trapped, he held it by its feet to show his friend. Then, in one swift movement, and entirely for my benefit, he momentarily held the bird against his friend's groin, throwing me a cheeky glance as he did so – the clearest signal that for them the red bird was equivalent to a penis.[22]

The Reverend Hugh MacPherson also noticed the association

between bird catching and sex, and illustrated it by using a particular painting as the frontispiece to his book *The History of Fowling* published in 1897. The painting was *Spring* by the French artist Nicolas Lancret (1690-1743), imitator and one-time pupil of Watteau, painted in 1738, and it provides a particularly explicit illustration of this link. The eighteenth century was a saucy and frivolous period among the French aristocracy whose amorous, aimless antics were immortalised by artists like Watteau, Fragonard and Lancret but were later brought to an abrupt halt by the Revolution. Lancret's picture shows two bird catchers, one playing his flute, the other about to pull on the clap net rope, and they are accompanied by no fewer than four gorgeous young women, each one egging the men on to catch song-birds for them. This is sexual selection in action. The two men are competing and, as MacPherson points out, keen to exhibit 'their skill as fowlers to the graceful dames whose smiles they strove so earnestly to win'. Lancret's painting has a double poignancy – catching wild birds to catch what they hope will be wild women. For the bird catch-ers success brought at least a twofold benefit: it demonstrated their skill and enabled them to offer a nuptial gift in the form of a songbird – much cherished by upper-class ladies.[23]

Göttingen to Bremen

Hans Duncker completed his book *Wanderzug der Vögel* (The Migration Routes of Birds) in March 1905. It so impressed Ehlers that he encouraged Duncker to apply for a Göttingen University prize, the Petsche-Labarrestiftung, which he duly won. Duncker was delighted and dedicated the book to his 'highly admired teacher, Ernst Ehlers on the occasion of his seventieth birthday'. His formal training now over and with a prize and doctorate in the same year, Duncker moved north to Bremen to begin his career as a high-school teacher. The Old Gymnasium was a classical state school and the best in Bremen, situated in a busy part of town adjoining its oldest district – the Schnoor, with its tiny medieval houses and narrow winding streets. Carved above its main entrance in big gold letters was the school's motto: *Ingenuarum artium studiis sacrum* – Dedicated to the Study of the Most Excellent Arts.

The Bremen to which Duncker moved in 1906 was a prosperous middle-class place. One of Germany's Hanseatic cities, its people were liberal, free and confident. Located on the Weser river, which ran broad and navigable a full fifty kilometres inland from Bremerhaven on the North Sea coast, Bremen was a busy, bustling port filled with the aroma of roasting coffee, one of the city's key imports. This was a civic society, in which commissions of honorary members took care of public and cultural life. Society ladies were identified by their tailor-made suits, pink nail polish and pearls, and season tickets to the opera. To thrive in Bremen one had to adopt the soft-spoken, formal manners of the local bourgeoisie. Duncker, with his three-piece suits, cigars and solemn academic confidence, seemed a perfect fit, although his restrained exterior concealed an inner dynamic enthusiasm for anything he decided to tackle. Anywhere else in Germany the title Herr Doktor, which under German law was now an official part of Duncker's name, would have given him a head start, but in Bremen, a city brim-full of wealthy merchants, consuls and businessmen, it offered no special privileges. The director of Bremen's psychiatric clinic summed up the situation by saying, 'the Bremen people keep some scientists as princes used to keep jesters'.

One of the first things Duncker did on moving to Bremen was to join its natural history society, based in the city museum. Built only ten years earlier, in 1896, the museum was located on the edge of the huge plaza outside the equally recent and architecturally magnificent railway station. The Übersee-Museum was a typical product of its age: a huge neoclassical building whose hard edges were softened by two voluptuous sphinxes guarding its entrance and a suite of anthropological and zoological reliefs around its first-floor windows. As a recent city guide explains, the museum was constructed specifically to house the numerous natural treasures 'found in or stolen from' Germany's many overseas colonies. Bremen wasn't alone in accumulating such artefacts and its museum was one of many erected across Europe and America as countries competed for the best displays of cultural plunder.

In 1907, after a year of military service, Duncker married Elsa Zwernsmann in Dessau, where they had been childhood sweethearts, and brought her with him back to Bremen. He then changed schools, moving to the Realschule am Doventor. In August the next year their first daughter, Marigrita – named after Duncker's mother – was born. In 1909 he changed schools again, moving to the Realgymnasium, an imposing boys' school on Hermann-Böse Strasse where he remained as a teacher of maths, physics and biology for the rest of his career. Keeping a pachydermal eye on the school from the other side of the road was an eccentric reminder of Germany's colonial ambitions: a life-size sculpture of a Namibian elephant made entirely from red bricks.

During his spare time in the years immediately before the Great War, Duncker produced a series of illustrated biology textbooks for schoolteachers. He also continued to pursue his fascination with birds and, using information from Göttingen's libraries and specimens from Bremen's museum, he explored the links between the geographical range and evolutionary history of different species. This was an ambitious project, but one that never came to much, and when it was published in 1912 Duncker was probably frustrated by its lack of clear conclusions. He was hungry for intellectual stimulation and longed for something that would engage his active brain, something novel, but something eminently doable. Before he could find anything the Great War intervened and any thoughts of studying birds or anything else were pushed to the back of his mind. Duncker was thirty-three. The family, which now included six-year-old daughter Marigrita, had just moved into a smart three-storey terrace house with an arched doorway and decorated roof in the south-east corner of town close to the Weser river. They lived in this house for the next thirteen years and it was here that their second child, Hans-Eberhard, was born and died less than one year later, and where their second daughter, Lotti, was born in August 1915. Serving as an officer on both the eastern and western fronts, Duncker gave a good account of himself, despite being slightly wounded; he was awarded several medals, including the Iron Cross. Four years later, with peace restored, Duncker settled back into his routine at the Realgymnasium. More worldly wise and picking up the pieces where he had left off, he continued as a conscientious, enthusiastic and innovative teacher, highly valued by both pupils and colleagues alike. This was a time for taking stock and Duncker looked at what had happened in biology elsewhere in the world while Germany and Britain had been preoccupied by a bloody, muddy war. Despite his commitment, teaching wasn't enough to occupy his active mind and he started to look again for an extra-curricular project, but it would be another three years before Reich's nightingale-canary caught his imagination.[24]

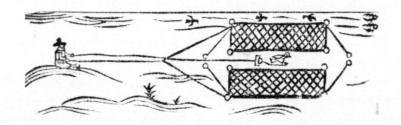

3

The Music of Nature

The Germans pay more attention to the song of these birds
[Canaries] than any other class of breeders, teaching them, when
young the notes of the nightingale and titlark; thereby adding much
pleasing harmony to the song of their birds.

THOMAS ANDREWES, *The Bird Keeper's Guide and Companion* (1830)

From the dawn of civilisation people have kept birds for their
song. 'Singing birds are so pleasant a part of the creation . . . they
were undoubtedly designed by the Great Author of Nature on
purpose to entertain and delight mankind,' wrote Eleazer Albin in
1737. People trapped and caged anything that sang: larks, thrushes,
wrens, warblers, goldcrests, robins, buntings and finches. In the 1300s
Italian gardens were festooned with caged songbirds to create an
auditory illusion of heaven. Singing birds were the medieval radio
and having a songbird in the house or garden was rather like having
piped music. Songbird enthusiasts were more discerning than those
who promulgate piped music, however, for they carefully selected
birds according to their vocal prowess and from the earliest times
(until the canary ousted it) the nightingale was the favourite with its
luscious, heart-stopping song.[1]

Writers and poets have endlessly extolled the nightingale's
vocal virtues, including the great German cage bird authority,
Johann Bechstein. In 1795 he wrote, 'The bystander is astonished

to hear a song, which is so sonorous as to make his ears tingle . . .' and the French natural history writer George-Louis Leclerc, Comte de Buffon wrote in the 1750s, 'The name of the Nightingale recalls to the memory of every man who has not lost the capacity of simple and natural enjoyment, the remembrance of some beautiful spring night, when the sky was clear, the air tranquil, and nature lay in expectant silence, as he listened enraptured to the songstress of the grove.'

Even the English loved nightingales and the seventeenth-century writer Isaak Walton declared that the bird 'breathes such sweet lowd music out of her instrumental throat that it might make mankind to think miracles are not ceased'. Many writers other than Buffon and Walton erroneously assumed that it was the female bird that sang, a myth perpetuated by Oscar Wilde in his children's story *The Nightingale and the Rose*.

The nightingale was all too easily deprived of its liberty, as a mid-nineteenth-century London bird trapper recalled, '[It] is a beautiful song bird; they're plucky birds too and answer to anybody when taken in April they are plucked [plucky] enough to sing in a cage. I can ketch a nightingale in less than five minutes: as soon as he calls I calls to him with my mouth and he'll answer me (both by night and by day) – I set my traps and catch 'em almost before I've tried my luck and I've ketched sometimes 30 in a day.'[2]

Nightingales were so popular they motivated generations of bird enthusiasts to devise ways of keeping them in good health and breeding them in captivity. The definitive guide was a French monograph, *Aedologie, ou Traite du Rossignol Franc, ou Chanteur* by Louis Daniel Arnault de Nobleville, published in 1751, which recommended tying the wings of newly caught birds so they couldn't batter themselves in their confined quarters. It also advised nurturing nightingales on raw heart (with the 'strings' removed) and ants' eggs (pupa), but even with all this attention, the death rate of newly caught birds was high. Those that survived the first weeks of incarceration, however, could live for several years and in the care of a handful of experts – mainly German – they could even be persuaded to breed.[3]

There's a contemporary joke which claims that in heaven all the policemen are English, the car mechanics are German, the cooks are French, the hotel keepers are Swiss and the lovers are Italian. But in hell the policemen are German, the car mechanics French, the cooks English, the hotel keepers Italian and the lovers Swiss. These national stereotypes are centuries old. The European aristocracy cared very much about their cage birds and were almost certainly telling jokes about their different attitudes to keeping them. From the sixteenth century onwards the Italians, French and English were renowned for their hedonistic obsession with cage birds, but were too idle to breed their own. The main interest the Swiss had in birds concerned the artificial ones that appeared from the inside of wooden clocks with impeccable if monotonous regularity. The Germans, on the other hand, were industrious and highly organised in trying to get their birds to breed in captivity, and they studied their charges with academic alacrity.

One of the most remarkable of these was the sixteenth-century German nobleman Freiherr Johann Ferdinand Adam von Pernau, Lord of Rosenau, and if the Karolinska judges ever decide to award Nobel Prizes to the long dead, they should give one to him. Pernau's methods of studying both wild and captive birds were remarkable and anticipated the Nobel laureate and animal behaviourist Konrad Lorenz by almost 400 years.[4] Virtually unknown during his lifetime and for centuries afterwards, Pernau was rediscovered and acclaimed by the great German ornithologist Erwin Stresemann only in 1925. This belated recognition occurred because, not wanting to 'hunt after honours', he had chosen to write anonymously, presenting himself merely as 'a fancier contemplating the creatures made by God'. His passion for birds arose, he said, as an antidote to the moral depravity of his time – gambling, shameful gluttony and what he euphemistically but accurately referred to as 'pleasure'. Pernau was unique in observing birds rather than eating them and his aim was to 'show how greatly

THE MUSIC OF NATURE

Man can delight in these lovely creatures of God without killing them'. He was also aware of the importance of making his own observations and keeping his discoveries distinct from what was known already. It was this individualistic approach that put Pernau ahead of his time.

He was among the first to appreciate that birds acquire their songs by learning, noting that if a young bird was brought up in isolation, unable to hear others, it 'never will attain its natural song completely, but will sing rather poorly'. But if instead of being reared alone, a young bird was brought up hearing only the song of a different species, it would learn that song. 'A finch', he wrote, 'will learn to imitate some of the Nightingale's strophes . . .' Pernau noticed as well that birds differed in their need to hear others singing. It was essential, for example, that a young chaffinch hear its own kind if it was to sing properly as an adult, but it was a waste of time using another species as a tutor since chaffinches could learn nothing but chaffinch song. At the other extreme, the simple song of the North American Alder and Willow Flycatchers, for example, is 'hard-wired' and young birds brought up in auditory isolation can sing just as well (or as badly) as any wild flycatcher. The canary, Pernau noted, lay at the chaffinch end of the spectrum, but with the remarkable capacity to transform virtually any sound, including a nightingale's song, into a melody.

Seventy years later and apparently unaware of Pernau's work, the Honourable Daines Barrington, an English lawyer, Fellow of the Royal Society and friend of Gilbert White, published in 1773 what he thought was the first scientific analysis of birdsong.[5] A wealthy dilettante whose intellectual pursuits straddled art and science, Barrington was keen to make sense of the natural world and felt that the best way of doing so was through the power of reason and the careful organisation of information. It was he who provided Gilbert White with a set of printed forms on which to tabulate his daily weather records, allowing White to link them to natural events such as the appearance of the first spring swallow or the call of the first cuckoo. The end result was White's *The Natural History of Selbourne*.

Of the many intellectual exercises Barrington carried out, his assessment of the quality of birdsong was among the best. He loved music and had previously evaluated the eleven-year-old Mozart when he visited England with his father in 1765. Barrington's genius with birdsong was to modify a method of assessment originally devised by the French art connoisseur Roger de Piles to compare the relative merits of colourists like Rubens with design enthusiasts like Poussin. The technique involved ranking paintings according to a number of criteria that Piles had dreamed up as he languished in a Dutch prison during the 1690s after being convicted of spying. Standing on Piles's shoulders, Barrington identified five criteria by which birdsong could be ranked and then, using a set of forms, set about scoring the vocalisations of seventeen commonly kept songbirds. Each was awarded a score out of twenty for: (i) mellowness of tone, (ii) sprightly notes, (iii) plaintive notes, (iv) compass (by which he meant overall range) and (v) execution. The nightingale, of course, came out on top, attaining 19 points for all qualities except 'sprightly notes', for which Barrington gave it only 14. 'I make 20 the point of absolute perfection,' he wrote, as reluctant as any schoolteacher to give his or her pupils full marks. Of 100 possible points, the nightingale got 90. Next in line was the linnet, scoring 74, then the skylark with 63 (faring badly on mellowness and plaintive notes). The reed bunting was at the bottom of the class with just 8 points. Barrington recognised that his system was subjective. 'I shall not be surprised', he said, 'if . . . many may disagree with me about particular birds.' This is exactly what happened.

Fifty years later Patrick Syme, a Scottish bird keeper and artist, was clearly irritated both by Barrington's scores and by his failure to define exactly what he meant by terms like 'mellowness' and 'sprightliness'. So Syme rescored all the birds in Barrington's original list as well as some others such as the canary,[6] pointing out those cases where he and Barrington disagreed and taking great care to be very clear about what *he* meant by 'mellowness' and 'sprightliness'. One of the birds on which they disagreed was the bullfinch, whose song Syme considered 'very sweet . . . soft and melodius'. But Syme was out on a limb here: Barrington hadn't even included the bullfinch in his list, because like virtually everyone else he felt that its song 'without instruction, is a most jarring and disagreeable noise'. While Barrington and Syme disagreed over a few species, in most cases their scores were fairly consistent.

Canary Scores

For some reason Barrington hadn't scored the canary, but Syme gave it 67 out of 100, placing it second after the nightingale in his scheme (the linnet got only 56). Few English bird keepers at that time would have taken issue with this ranking, but had this exercise taken place in Germany, the canary would probably have received a much higher score and might even have jostled the nightingale for first place. The entire focus of canary breeding in Germany was on song, whereas in Britain song was secondary to posture, 'type' and overall appearance. For over 200 years through a combination of selective breeding and intensive training German bird breeders so modified and improved the bird's song that by the late 1700s some canaries were said to sing as sweetly as any nightingale.[7]

Despite the nightingale's superiority as a singer, it had several failings. These shortcomings were eloquently outlined in the early 1700s by Monsieur Hervieux who, although he had a vested interest in promoting the canary, was startlingly honest in his assessment of nightingales.[8] They were difficult to keep, he said, and needed a special diet that required 'much Application'. 'Besides, the Nightingale, after all the pains taken in feeding and rearing sings but one short Season of the Year'. Nightingales died all too readily in captivity partly because every autumn they went through several weeks of hormonal hell. Battering themselves against the bars of their cage, they acted out their entire migratory journey in the frantic but futile motions that German researchers refer to as *Zungunruhe*.

Canaries, on the other hand, were much more robust. They didn't migrate, they were more easily maintained and bred in captivity, and they would sing throughout the year. Paradoxically, perhaps the canary's greatest asset was that its song was not perfect yet could be made so with the right training, an activity that provided a challenge for the trainer and the bird itself. Of the thousands of bird-fanciers who tried to coerce canaries into singing the nightingale's song, Karl Reich was the only one to produce an entire strain of birds that could do so.

Canary Mania

On hearing Reich's nightingale-canary for the first time in 1921, Hans Duncker was inspired to spend the next fifteen years of his life studying canaries. To understand why he did so, we must go back in time to when they were first discovered.

It is said that when Sir Walter Raleigh returned from his travels in the 1580s, one of the first things he did was to present Queen Elizabeth I with a cageful of wild canaries. She was singularly unimpressed. Far from being bright yellow, the canary was then an insignificant dull green little bird. Sir Walter begged her not to judge the birds by their appearance but to wait until she had heard them sing.

When they performed in due course, she was overwhelmed by the vigour and variety of the birds' voices – a song to challenge the nightingale's in quality and compass.[9]

In fact, on hearing a canary sing for the first time, almost everyone was astonished by it. Conrad Gessner,[10] a Swiss naturalist and polymath whose animal encyclopaedia was published during Elizabeth's reign, described the canary's song thus:

It hath a very sweet and shrill note, which at one breath continued for a very long time without intermission, it can draw out sometimes in length, sometimes raise very high, by a various and almost musical inflexion of its voice, making very pleasant and artificial melody. The sound it makes is very sharp, and so quavering, that sometimes when it stretches and exercises its little throat and Chaps [mandibles/jaws], whistling with all its force, it vehemently strikes, and even deafens the Ears of the hearers with its shrillness. Many are delighted by this kind of singing, many also are offended, saying, that they are astonied [astounded or disconcerted] and deafned by it.

Why do canaries sing so loudly? We don't know for sure. But stuck out in the Atlantic, the Canary Islands are extremely windy, dissipating sound like pollen from grass, and so it is possible that they sing louder than just about any other small bird, simply in order to be heard. Remarkably, some of the indigenous people of the Canary Islands, on La Gomera in particular, also communicate by whistling. Silbo Gomera is a whistled version of spoken Spanish, and is used not because of the wind, but as an efficient means of communicating across the vast ravines and gullies that intersect the island. This whistling language once allowed the indigenous people to communicate among themselves without divulging any secrets to their unwelcome Spanish masters.

Canaries reached central Europe via Spain and Italy in the 1400s and on a continent of bird keepers their remarkable song immediately captured people's attention. The Italian nobility were completely infatuated and imported them in hundreds. But the Germans were among the first to devise clever ways of propagating them in captivity. Despite centuries of bird keeping, there had never been any need to breed birds in captivity because they were readily available in the wild. But canaries were so esteemed, so expensive and so difficult to acquire that there was a tremendous incentive to create a self-perpetuating, captive population. This was not a simple undertaking.

Canaries were not common pigeons or fowl, but more like feathered orchids demanding extraordinary care and skill. The closest that previous generations of bird-fanciers had come to propagating their own birds was rearing young ones taken from nests in the wild. Once wild-caught canaries started to lay eggs in captivity the temptation was to rear their babies by hand. Undoubtedly this was tried but it failed because, rather like captive chimpanzees brought up by humans, hand-reared birds rarely reproduce properly as adults. During the rearing process they become sexually fixated on their owner rather than their own species. The Germans succeeded by giving their wild-caught canaries lots of space, usually a large room planted with small trees, re-creating as far as was possible a micro-cosm of their natural environment. By the mid-1600s the German fanciers were exporting canaries across Europe in such large numbers that they became known as German-birds.[11]

The obsession with canaries and other singing birds at this time provided great opportunities for anyone who could increase the quality or quantity of song produced by a captive bird. Early bird breeders recognised that canaries and other songbirds, such as linnets and chaffinches, did not hatch with their songs already embedded in their brains, but learned them from older birds. In captivity young birds could be taught either by their own species or a different one, by a human whistling or even by a musical instrument. The importance of learning in the canary's acquisition of song was beautifully, if acci-dentally, demonstrated when one particular bird in the 1700s incorporated the sound of distant church bells into its repertoire. Another, owned by a tax collector, included an unusual 'clink' in its song which turned out to be the 'telling of crowns'.[12]

German-birds were soon considered far superior singers to those taken from the wild and were very much in demand. As Barrington wrote, 'Most of those Canary-birds, which are imported from the Tyrol [then part of Germany] have been educated by parents, the progenitor of which was instructed by a nightingale; our English

Canary-birds have commonly more of the tit-lark note. . . . The traffick in these birds makes a small article of commerce, as four Tyroleze generally bring over to England sixteen hundred every year; and though they carry them on their backs one thousand miles, as well as pay £20 duty for such a number, yet upon the whole it answers to sell these birds at 5d, a piece.' Barrington probably didn't know what these canary traders went through to safeguard their cargo. The main worry was disease, which could spread like wildfire through a trader's entire stock.[13] To minimise the risks of infection, every few days the canary salesmen checked into an inn, hiring a number of rooms in which they released their entire stock of several thousand birds while they scrubbed out their cages. Once the cages were clean and dry, the birds were recaptured and replaced, and the trader moved on, presumably leaving somebody else to clean out the rooms.

By the time Barrington was writing about birdsong, the centre of canary breeding had shifted north from the Tyrol to Nuremberg and Augsburg, and these two towns dominated the business for the next century. It shifted again in the early 1800s when miners from the Imst Valley brought canaries north to the tiny mining town of St Andreasberg in the Harz Mountains, midway between Berlin and Frankfurt. Mining towns in this region had a long tradition of keeping songbirds, especially chaffinches, which were trained to sing for competitions. It is not difficult to imagine the enthusiasm with which they greeted the much more malleable canary and St Andreasberg quickly became the heart of canary breeding in Germany.

In the early 1920s, when Reich and Duncker were planning their first experiments, St Andreasberg was the main organ pumping roller canaries around the entire world. Almost everyone in the village, man, woman and child, was employed in some aspect of this remarkable business – but only in their spare time, for the main employment here was mining. Tucked away on the western edge of the Harz, the village of St Andreasberg grew up around the Samson silver mine, which first opened in 1521.[14] Conditions underground were appalling and the miner's life a harsh one – very few reached old age. The main lode was a kilometre below ground and the workers descended to it on an ingenious 'man-engine' comprising two adjacent vertical rods pumped alternately up and down. This enabled a man to descend (and ascend) fairly effortlessly, by stepping first to one side and then to the other – albeit for a full forty-five minutes. Once they got to the main vein it was hellish: 40°C and 100 per cent humidity. The air was often gaspingly poor in oxygen or fatally rich in carbon monoxide, particularly at the ends of the long galleries. Generations of miners had protected themselves from bad air underground by taking small birds with them. Mice had been tried, but they proved far too robust, a fact confirmed by experiments conducted in the early 1900s which showed that mice could tolerate nearly a hundred times as much carbon monoxide as a canary before showing any warning signs. Small birds were supremely sensitive to poisonous gas and the canary provided the added benefit of its voice, which must have sounded eerily reassuring as it reverberated around the ghostly galleries.

Following its inception in the early 1800s, the St Andreasberg canary business expanded rapidly and by the 1820s was producing about 4000 songsters each year. Other Harz villages were also busy rearing birds, but the tipping point came in 1836 when a survey declared St Andreasberg canaries the best songsters in Germany. In the canary mania that followed, across Europe and as far away as Russia and the United States, people wanted the genuine article – a Harzer roller. A combination of genetic good fortune and skilful training had generated a bird with a distinctive and attractive song, and breeders could get as much as 100DM for a good singer. By 1882 three-quarters of the 800 families in St Andreasberg were rearing canaries and in some years the town exported as many as 12,000 male birds. The Harz region as a whole exported a phenomenal 150,000 males annually, providing a good living for a few enterprising bird dealers.

This remarkable business also generated subsidiary jobs such as cage making. Every bird that left the Harz did so in its own individual cage, put together by families working together during the long winter evenings. Children were a crucial part of this cottage industry, their small fingers being particularly adept at fitting the tiny wooden pieces together. And the cages themselves were tiny, allowing the bird only just enough room to turn. In design they were essentially the same as those used by the Tyrolese traders who had lugged canaries across Europe on their backs two centuries earlier. The cage parts – slats and dowels – were provided by the sawmill on the edge of town. Families assembled them, returned them as completed cages and were paid just two pfennigs per cage for their efforts. The other subsidiary job was making the diminutive white ceramic pots that went into every cage to hold the canary's drinking water.

The Harzer or roller canary's reputation rested on its song, so called because it resembled a deep, hollow 'roll' – a repetitive 'rorororo' which sounded as though it was sung from deep inside a barrel.[15] Totally unlike the warbling, whistling wild canary, the roller's song is pitched much lower, highly repetitive and, to my ear at least, far from melodic (Figure 1).

On the other hand, the roller's song is relatively loud and curiously sustained, and perhaps it was the novelty of the sound that people found captivating. Despite its monumental success, not all canary enthusiasts were equally enamoured of the roller, and during the early 1700s a breakaway group of enthusiasts from the town of Malines (Mechelen) just south of Antwerp produced a more melodic version known as the 'water slager' or Malinois. 'Water slager' means 'water-beater' and these birds were so named because the main feature of their song is a tongue-clicking 'klok', like a large drop of liquid falling into a barrel of water. The song also contains other bubbling sounds, and is generally much more varied and exuberant than the roller's restricted repertoire.[16]

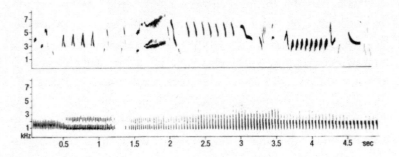

FIGURE 1 *Sound sonograms of the songs of a wild canary (upper) and a roller canary, showing the clear difference in the melodic complexity of their songs (from Güttinger, 1985).*

Canaries require several weeks of training before they are proficient singers and it was usually winter before the young rollers were ready for sale. Hundreds of cages, each containing a single bird, were loaded on to horse-drawn wagons in the villages and driven to Göttingen. They were then placed on the train for Bremen, where they were put on the Southampton steamers and sent across the sea to New York. The American lust for canaries seemed insatiable and in the first four decades of the twentieth century, before World War II interrupted the trade, the USA imported over 10 million of them.

The boom in Harzer canaries lasted from the 1870s until about 1910. Its demise is said to have occurred for exactly the same reason as the decline in whales, cod and seahorses – greed. During the period of maximum demand the competition between dealers became so intense that, desperate not to miss the boat, some of them foolishly sold birds before their training was complete and these inevitably poor singers killed the market. Without doubt other factors also contributed to the canary's decline, including a glutted market and changing fashions in birds and other commodities.[17]

A Bear's Melody

Karl Reich was well known among Germany's canary cognoscenti for his recordings of birdsong and for his nightingale-canaries. But Duncker knew virtually nothing about him and when he knocked on Reich's door that summer afternoon he could have had little idea how their meeting would shape the rest of his life.

Four years Duncker's junior, Reich ran a family hardware business on Fedelhörenstrasse. Tall, slim and with eyes that twinkled behind his spectacles, Karl Reich was the archetypal shopkeeper, radiating enthusiasm for both his work and his hobby. Duncker went to Reich's home on Am Wall, a street in the city's original fortifications, now overlooking the Wallanlagen park separating the old and new Bremen.[18] There is no record of exactly what happened during that initial meeting, but based on my own experiences with bird keepers, and the way Reich's and Duncker's friendship blossomed, this is what I imagine took place.

Welcomed by a smiling Frau Reich, Duncker was ushered through into her husband's bird room. This was Reich's den, his potting shed, his private sanctum where he could devote himself to his birds. After the two men greeted each other and shook hands, Duncker must have needed a moment or two to take it all in. On one wall was a massive collection of rosettes, cups and canary-motif vases signifying Reich's success. On the other walls were row upon row of elegant wooden cages, each with a pair of solid doors, most of which were firmly shut, but nonetheless containing birds. The cages matched the dark, polished wooden furniture in the room, each cage bearing on its door a golden marquetry canary emblem. Then there was the warm and rather pleasant scent of the birds and their aromatic seed diet – a smell that undoubtedly reminded Hans of his childhood days when he'd visited his neighbours' bird rooms. There was also the sight and sound of the birds themselves. No sooner had Reich opened the doors of a particular cage than the occupant burst into song. This was no ordinary song, but an enchanting, sonorous, reverberating, rolling chorus, so powerful that Duncker could feel it in his chest. Reich loved the effect his birds had on people. As he opened the doors of cage after cage the birds' song built up to a crescendo – a veritable orgy of sound.

After five minutes Reich started to close the cage doors one by one, silencing the birds in turn. The first part of the performance was over and as he closed the last door the silence was uncanny. Reich waited until the sound of the singing had ceased ringing in their ears and then without speaking cast a glance at Duncker, as if to say 'Now listen to this' and opened the doors of another cage. Inside was a single green bird which immediately began to sing, remarkably, Duncker noticed, with its beak firmly closed. The sound that the bird uttered was not a typical roller canary song but the gorgeous, liquid notes of a nightingale. Duncker was transfixed, realising immediately that this was the song he had heard in the street the week before. The little bird held its body rigid at a low angle on the perch and literally shook with the effort of its powerful song. Duncker was visibly moved and, closing his eyes, allowed himself to be transported back for a moment or two to the Ballenstedt woods and the wild nightingales of his childhood. Reich beamed, for this was exactly the way he liked to see people react to his birds. He was even happier when Duncker told him the story of the mysterious 'nightingale' he had heard in town.

Almost as if she knew the routine by heart, once the nightingale performance was over Frau Reich came in and left a jug of beer and two glasses on the table. Reich told Duncker the story of how he had created his extraordinary birds. It started when, as a small boy, he had been given a canary as a Christmas gift. He loved canaries and revelled in their rich baritone voices, but he didn't much like the modern Harz Mountain rollers with their relentlessly repetitive songs. Instead, he longed for something more satisfying and in 1909 set about trying to find the best song tutor for his canaries. He tested no fewer than twenty-five different bird species from around the world, before deciding, like Pernau and Barrington before him, that he could do no better than the local nightingale. Then, in 1911, when he was twenty-six, he produced a canary with a voice of truly outstanding quality, even by roller standards. This bird, which he named *Bär* (Bear), changed his life. He had a voice so deep and so pure, Reich said, that it reminded him of the rich, liquid tones uttered by nightingales and inspired him to try to create an entire strain of birds that would sing like this. It was too late for Bear, who sang a typical canary song, but Reich knew that with the right training Bear's offspring might be induced to sing the ultimate tune. The following year he mated Bear back to his mother, then placed their offspring, at the tender age of three weeks, under the care of their 'schoolmaster', a vigorously singing nightingale. It started off well, but no sooner had the young canaries started to warble their first liquid notes than the tutor stopped singing. The canary's breeding season, which began in May, coincided with the end of the nightingale's singing season. Bear's male babies eventually uttered a smattering of nightingale strophes interspersed with typical canary phrases – hardly a success, but a result sufficiently tantalising to spur Reich on to finding a solution to the nightingale problem. He would have to shift the nightingale's singing season so that it encompassed the young canaries' learning window. This meant getting nightingales to continue pouring out their wonderful song long into June and July – months after they would normally have stopped singing. Remarkably, Reich told Duncker, he was able to do just that by modifying the nightingales' diet and the temperature at which they were kept so that they moulted and came into breeding condition later than normal.[19]

I admit that I was incredulous when I first read of Reich's claim, in an old interview, that he could alter the timing of his nightingales' song in this way. A bird's singing season is deeply entrained and generations of bird keepers have succeeded in changing the timing of birds' moult and singing season *only* by altering the amount of light they received. Reducing the light a bird received in May – 'stopping', they called it – fooled it into thinking the season further advanced than it really was. This meant they moulted sooner and instead of starting to sing in spring, they were going full belt by September or October.[20] When asked, Reich told everyone including Hans Duncker that he had discovered the temperature-and-diet trick from a very old anonymously written German book with a long-winded title: *Unterricht von den verschiedenen Arten der Kanarievögel und der Nachtigallen, wie diese beyderley Vögel aufzuziehen und mit Nutzen so zu paaren seien, dass man schöne Zunge von ihnen haben kann* (Lessons about the different species of canaries and nightingales, how these two birds should be bred and usefully mated to produce a nice tongue [*sic*] from them). The book certainly existed (it was published in 1772) and it is very rare, but on tracking it down I found no mention of any technique other than 'stopping' to alter the birds' singing season. Reich had been deliberately deceptive, as indeed had the anonymous author of his ancient book for the crucial chapter on training nightingales to sing throughout the year was lifted word for word, with no credit, from the book by Arnault de Nobleville – the French authority on nightingales – published twenty years earlier. It is incredible that no one bothered to check what these early authors had said and that no one rumbled Reich's secret. But they didn't, presumably because these books were as rare then as they are now and resided only in the hands of a few avicultural bibliophiles.

Reich prided himself on being able to produce singing nightingales at almost any time of year, but apart from the temperature-and-diet story, he was always evasive about the methods behind his success. He later claimed that he refused to reveal his secret because he didn't want everyone rushing out to catch wild nightingales, but the fact was that, just as with his hardware business, he didn't want any competition, for his nightingale-canaries were unique and justifiably famous.

With a succession of nightingale 'schoolmasters' for his young canaries, Reich selected and perpetuated his precious stock with all the skill of a master breeder. Both Reich and his birds excelled. One of his stud males – a descendant of Bear – was paired a staggering thirty times in one season, and produced no fewer than seventy-two offspring. The bird paid the ultimate price for his lifestyle and died soon after the season ended. He was duly stuffed for posterity as were all the birds in the Bear dynasty. Still, Reich's management skills were such that he could produce as many as 350 young birds in a season. Every male he used was one of Bear's descendants and they all bore his vocal qualities; the females were Bear's relatives too – rather more distant – the song was in the birds' blood. Within a few years Reich was able to dispense with the nightingale tutors altogether, for his young canaries now acquired their unique song entirely on their own. Others, like Baron von Pernau and Arnault de Nobleville, had produced one-off nightingale-canaries, but no one had ever created a self-perpetuating strain of them.

Since 1910, Reich told Duncker, he had made phonograph recordings of his birds on shellac discs and sent them to friends. Reich was the pioneering maestro of birdsong recordings, training his birds to sing to order in front of the cumbersome recording equipment – no mean feat and absolutely essential in those days before editing. He must have had a truly magical touch, for he persuaded one particular nightingale to perch and sing right *inside* the horn of the recording machine, thereby ensuring spectacularly clear recordings. Reich even used his records of nightingale songs to train his canaries – the first time this had ever been done and the first evidence that sound alone was sufficient for a bird to learn a song. In 1910, just three weeks after he made his first phonograph record, he was invited to describe his pioneering techniques and play his nightingale recordings at an international bird conference in Berlin. In the darkened auditorium the audience listened with rapt attention to this dual miracle of modern technology and evolution.[21] His gramophone records made him famous and for the next twenty years Reich continued to make recordings of birdsong, which were sold across Europe, Russia and in the United States. Although they made money, Reich said it was the kudos he enjoyed most.

The Status of Song

Duncker left Reich's home with a million ideas racing through his head. As the weeks went by those ideas wouldn't go away and he became increasingly intrigued by what he had seen and heard. As well as exciting scientific interest, the nightingale-canaries provided a welcome distraction from the degenerating political and economic situation in which Germany found itself following the Great War. The Allies' peace terms and voracious reparation programmes were driving the country into turmoil. By the early 1920s rocketing infla-tion, mass unemployment, malnutrition and terrorism had generated an atmosphere of despondency bordering on desperation among the German people, especially the middle classes. Perhaps Duncker latched on to Reich's canaries as someone in a shipwreck latches on to a piece of flotsam. In any case Reich's account left Duncker keen to place these special birds in the overall context of song-canary culture. After much reading and hours of discussion with Reich, he started to sketch out an evolutionary tree of canaries, similar to one Darwin had produced fifty years earlier for pigeons, and he placed Reich's birds on one of the outermost branches. He thus signalled that these were birds whose song was most improved over the original wild ancestor, the most advanced and the most successful product of artificial selection, and therefore the highest point in canary culture. There were two reasons why Reich and his German predecessors were so motivated to create such birds. Most obvious was the beauty of the birds' song. But it wasn't a question of simply satisfying some Teutonic aesthetic fantasy. Duncker, who had read his Darwin, must have seen the remarkable parallel between canary culture and Darwin's analysis of pigeon breeding: 'The action of unconscious selection, as far as pigeons are concerned depends on a universal principle in human nature, namely on our rivalry, and desire to outdo our neighbours. We see this in every fleeting fashion.'[22]

Male rivalry takes a multitude of forms and it was inevitable that bird keepers would eventually appropriate the displays which male canaries and other birds used to enhance their own status and to lure female partners, and for exactly the same reason. The first such contests were singing competitions, which were judged on either the quantity or the quality of the birds' vocal attributes. The species most often used were skylarks, goldfinches, linnets, greenfinches, chaffinches and later, of course, canaries.

The earliest record of these contests dates from 1456, when a chaffinch singing competition was held in the Harz Mountains. Singing contests may have started earlier still, when Otto the Great, Holy Roman Emperor, moved people into the Harz region to mine silver around 960, but the poor, unlike the rich, left few records of their pastimes.[23] Flanders also has a long history of chaffinch contests, dating back to at least 1593 when there was a competition at Ypres. For the 200 years prior to 1800, Flemish chaffinch contests took place only in cities because, perversely, only city dwellers were allowed to catch birds. Their rural relatives were banned from doing so by their landlords, who jealously guarded all hunting rights.

The traditional chaffinch contest was a 'strong singing' test and the winner was simply the one that sang the highest number of complete songs in a set period of time – about five minutes' duration. Another type, 'distance singing', was based on the number of songs a chaffinch performed in thirty minutes or an hour. In *Germinal*, Emile Zola describes an hour-long contest in a poor mining community in northern France. Fifteen nail makers, each with a dozen chaffinches – 180 birds in total hanging up on a fence in a yard – sing furiously against each other, each one urged on by its owner shouting at it in Walloon to sing more and more. A crowd of over a hundred spectators watches and waits until the score keepers eventually declare a winner and the owner receives his prize – a metal coffee pot.

In all types of competition the birds were kept in cloth-covered cages so that they could hear but not see their rivals – the same reason that Reich and all roller canary breeders kept their birds behind closed cage doors. Sometimes, as in Zola's story, the chaffinches were blinded – usually by the unspeakably cruel process of touching the eyelids with a heated wire – because without distraction the birds were believed to sing better. In Belgium, a hot spot for such contests, blinding became illegal in 1921.

Song contest chaffinches were always taken from the wild and trappers sought out birds with particular vocal attributes. Wild chaffinches typically sing in bursts lasting about two seconds, interspersed with rest periods of seven to fifteen seconds. Dramatic regional differences in chaffinch singing meant that trappers seeking 'distance singers' inevitably converged on those areas where birds sang with the shortest breaks. Among wild birds the duration of individual songs varies rather little, so in order to be a winner a bird must have extremely short rest periods. Cramming 600 songs into an hour means that a bird with a two-second song must rest for an average of no more than four seconds between songs. This is an extraordinary output. Yet the birds that take part in the competitions, which still flourish in Flanders, sing at an even higher rate – one exceptional bird recently poured out some 1400 songs during a sixty-minute contest.[24]

The passion for these contests among the chaffinch song subculture was, and still is, remarkable. In London towards the end of the nineteenth century, champion chaffinches had cult followings, rather like those of racehorses today. Unlike the Continental competitions, those in London pubs often comprised pair-wise tournaments. An account from the 1890s describes one such contest in the Cock and Bottle pub in London's East End. The match was between two 'stunners', Shoreditch Bobby and Kingsland Roarer, who arrived incognito in small, covered cages. As the clock struck eight the cages were hung side by side on the wall and their covers removed. As one bird started to sing the other responded, exactly as they would in the wild in a territorial dispute. The main difference here was that, unable to see each other, neither bird was intimidated and they continued to sing against each other relentlessly. Using a piece of chalk, two men, known as the 'markers', recorded every completed 'strophe' or 'limb' sung by each chaffinch. Occasionally the birds paused for a drink, much to the consternation of the crowd, for it was forbidden to shout or whistle encouragement, although coughing – which was unavoid-able – was allowed. Two minutes from the end of the fifteen-minute contest, the Roarer – ahead by twenty points – inexplicably stopped singing and started to feed! His owner, seemingly unperturbed, pulled out a red handkerchief to mop his brow and, as if by magic, the bird hopped back on to its perch, resumed its relentless outpouring and won the competition.[25]

Sustained singing is energetically draining, which is, of course, the whole point. The crucial question, for me, is whether in the wild these winning birds would also be the most attractive to females, both for long-term relationships and for extramarital sex, which is common among chaffinches.[26] My guess is that they would be. No one has meas-ured how much effort it takes for chaffinches to perform like this, but there is no doubt that these birds sing much faster than they ever would in the wild. That chaffinches occasionally dropped dead during singing matches in the past also says something about how demanding these contests were, although the birds' stress was undoubtedly exacerbated by the smoky pub atmosphere. The human equivalent of a chaffinch song contest is probably an operatic aria, the longest and most demand-ing of which is Brünnhilde's sacrificial scene in Wagner's *Götterdämmerung*. Like a chaffinch keeper, Wagner also seems to have pushed his performers to their limit. His first Tristan, Ludwig Schnorr, expired soon after the first performance. Some clarinet players say that ten minutes of continuous playing is about their limit, and 'Jesu Joy of Man's Desiring' is a particularly exhausting piece for oboe players. In men and birds alike, the performers run out of breath because their abdominal muscles become exhausted.

In Britain singing matches between chaffinches or other small birds, with the exception of the canary, are a thing of the past, although bird keepers still gossip about gypsies holding clandestine contests with £1000 stakes for goldfinches and other birds 'somewhere in Essex'. In parts of Germany and Flemish-speaking areas of Belgium, however, chaffinch song contests continue undiminished. In 1973 Belgium banned the catching of all small birds except for the chaffinch and in 1999 no fewer than 10,000 wild chaffinches, mainly migrating birds from Scandinavia and Russia, were legally taken during the October – November trapping period specifically for the song contest cult. The birds have to be trained to sing in the proper manner because their foreign dialect won't do at all. The competition rules are explicit in that the birds must finish each song with a Flemish flourish or *suskewiet*, and they must be taught to do this. The birds also have to be trained to sing very short songs. This is achieved by keeping freshly caught chaffinches with experienced birds that sing exactly the right way, in the hope that the new birds will pick up their tutors' singing style. This works with about half the birds; those that fail to master the new song are released. To produce a fully trained bird takes three years, which may seem a long time, but chaffinches are durable and successful birds may compete over ten or more years – the record is twenty-nine years. 'Fincheneers', as they are known, are notoriously besotted by their birds, spending more time with them than with their wives and shedding more tears over the death of a bird than they do over a relative.[27]

Chaffinch singing contests have lately been pushed to the brink of extinction by the animal rights movement. Because it will soon be forbidden to take chaffinches from the wild the prospects for the Flemish fincheneers are uncertain. If they are to continue they will have to use chaffinches bred in captivity and they are full of gloom at the prospect. Catching the birds is part of the fun, but they also feel that captive-bred birds will limit what they have to work with. Still, the native bird-keeping cult in Britain (which was interested only in the birds' appearance) suffered similar strictures fifty years ago and survived, just.

As soon as canaries fluttered down from their royal pedestal and became common and cheap, ordinary men started to pit them against each other in competitions. Initially, canary competitions were like most chaffinch contests and based exclusively on the quantity of song. But the last record of such contests is from Lancashire, northern England, in the late 1800s.[28] Competing by numbers quickly gave way to a more sophisticated competition based on quality, fostered by the canary's intellectual ability and inherently more varied song. This was possible because, unlike chaffinches and other songbirds, canaries were reared in captivity and so could be selectively bred and trained for the quality of their song.

4

Music in the Brain

It is love, and especially the first season of love, which inspires the song of these birds. The spring awakens in them the desire of love and the desire of song at once; the males are the most ardent and . . . if placed . . . where they are unable to satisfy their sexual desires, and so extinguish at the same time the love of song, will sing throughout the greater part of the year.

JOHANN BECHSTEIN, *Natural History of Cage Birds* (1795)

Since Aristotle's time it has been known that singing is closely linked with reproduction: that only male birds sing and they do so to announce their occupation of a breeding territory. In Darwin's theory of sexual selection birdsong was an expression of the competition between males for territory and the opportunity to breed. But, more contentiously, Darwin also suggested that song had evolved because females were charmed and attracted by it. The songs of male birds were shaped by female preferences, precisely because females chose to reproduce with particular singers. Darwin supported his view by quoting from the great German bird authority Johann Bechstein, who wrote, 'female canaries always select the best singer, and that in a state of nature the female finch selects the male out of a hundred whose notes please her most'. But what constitutes the 'best' song? Darwin assumed that the melodic nature of song, its degree of elaboration and complexity were what counted, much as they do for human listeners. Indeed, I suspect that the reason we like birdsong is precisely because it lights up the same parts of our brain that music does. Obviously, among birds what females find attractive varies from species to species. In some cases it is the melodic quality of song itself, in others it is the size of the song repertoire, in others still it is the rate or volume at which songs are uttered that matters.[1]

Confirmation of Bechstein's statement about female canaries preferring good songs was a long time coming. But in 1976, by means of some elegant experiments, Don Kroodsma at Rockefeller University New York showed how isolated female canaries built their nests more quickly and laid more eggs when listening to recordings of rich song repertoires than they did when listening to poor song repertoires.[2] The degree of elaboration of a song repertoire was clearly one of the cues females used to assess males. It has since been discovered that many animals, including ourselves, also prefer musical tones with a certain order and rhythm, in much the same way that our brains seem to prefer symmetrical over asymmetrical patterns. Kroodsma's experiments showed convincingly that females can and do discriminate between the vocal displays of different males, thus refuting the concerns of many of Darwin's sexist critics that females did not have the cognitive ability to make informed choices about their partner.

When Darwin turned to discuss humans in his book *The Descent of Man and Selection in Relation to Sex*, he started with a conundrum: 'As neither the enjoyment nor the capacity of producing musical notes are faculties of the least use to man in reference to his daily habits of life, they must be ranked amongst the most mysterious with which he is endowed.' But by referring to music as a mystery Darwin was being disingenuous. Just like other 'faculties' that appear to be of no use in daily life, music, he argues, is a sexually selected trait important in the acquisition of partners, exactly as it is in animals.

Darwin also noted that music predominantly arouses pleasant emotions, especially those associated with love, lust and triumph, rarely horror, rage or fear. The parallel with animals is clear: 'All these facts with respect to music . . . become intelligible to a certain extent, if we may assume that musical tones and rhythm were used by our half-human ancestors, during the season of courtship, when animals of all kinds are excited not only by love, but by the strong passions of jealousy, rivalry and triumph.' Music serves as a male sexual signal, exactly as song does for birds, with females preferring the best composer.

Darwin made his concept of sexual selection palatable and digestible for his readers with an analogy they could understand: artificial selection. Sexual selection was just like men choosing the most attractive pigeons or the largest cattle from which to breed. By doing so they perpetuated the desired traits. Female choice was therefore responsible for extreme male displays like the peacock's tail, the bowerbird's bower and the nightingale's song.

As Darwin was keen to point out, a crucial difference between man's artificial selection and sexual selection in nature was that sexual selection was usually kept in check by survival value. A peacock's tail is two rather than three metres long because of natural selection. Peacocks with tails longer than two metres might be monumentally attractive but would be hopelessly vulnerable to predators. Two metres marks the point beyond which the disadvantages of a large tail outweigh – literally – any reproductive benefit: tail length has been kept in check by natural selection. Artificial selection, on the other hand, knows no such bounds and in principle it should be possible to breed peacocks with even longer tails. The Japanese have done just this with the domestic fowl and produced the Onagadori or Phoenix strain, whose tail feathers trail up to five metres. Think, too, of all those portraits of strangely shaped prize-winning cattle with huge haunches and tiny heads. Darwin wrote, '. . . an attempt was once made in Yorkshire to breed cattle with enormous buttocks [which were no doubt extremely attractive to farmers and butchers alike] but the cows perished so often in bringing forth their calves, that the attempt had to be given up.' A similar problem occurred among the pigeons Darwin admired most, short-faced almond tumblers. This breed would not have survived in the wild, for their stubby beaks are too short for the young to break out of the egg unaided. German song canaries, including Karl Reich's strain of canaries that sang like nightingales, were similarly a sexual disaster from the canary's own standpoint. The problem, for the female canary, was that while the male had been artificially selected to display in a particular way, the female's preference had been left untouched. A male canary singing a nightingale song was about as sexy to a female canary as a male gorilla is to a woman – King Kong notwithstanding.

The wild canary's song, like that of other birds, is a product of sexual selection. In the past, male canaries that sang elaborate songs left more descendants because their song, as Bechstein noted, made them more attractive to females and thus gave them a competitive edge. On the face of it the creation of the roller canary looks like a man-made version of the female choice part of sexual selection: men chose which songs they preferred and allowed the best singers to breed. But this is an illusion. The features of canary songs that were attractive to men turn out to be decidedly unattractive to female canaries. The male canary's song is like a strand of DNA, comprising a mixture of crucial coding regions, spacers (or non-coding regions) and junk. The breeders of roller canaries naively focused their selection on the 'junk' part of the canary's song. The roller's 'rorororor' song is junk because female canaries are not turned on by it as they are by the songs of any other type of canary. Tucked away in the complex array of sounds that make up the wild canary's melodic and varied song are some very specific phrases or trills – equivalent to the crucial coding regions in a twist of DNA. It is these that really excite females, instantly inducing them to adopt a mating position. (The discovery of something similar among humans doesn't bear thinking about.) If these phrases are cut and spliced from a tape recording and played to a female canary, they have an almost immediate sexual effect, even in the absence of a male bird. The songs of roller canaries, on the other hand, are utterly devoid of these sexy syllables – so not only have breeders exaggerated the unattractive part of the canary's song, they have simultaneously eliminated the attractive bits.[3]

Birdsongs are generated in much the same way as blowing across a blade of grass held between your thumbs. Utilising the same principle Erasmus Darwin, grandfather of Charles, used a silk ribbon stretched between two pieces of wood and a set of bellows to create a sensational speaking machine in the 1770s.[4] Reed instruments like the clarinet also work by passing air over a membrane and causing it to vibrate. In birds the sound-producing membranes reside in a structure known as the syrinx, which lies in the upper breast where the windpipe divides in two to join the lungs. In humans these membranes, the vocal cords, are much closer to the mouth.

The grass-and-thumbs system for making noise is very crude, but if you've tried it you will know that you can easily modify the quality and volume of sound by altering the thickness of the grass membrane, by how tightly you hold the grass or by how hard you blow. Now imagine a similar system but with infinitely more control and you've got a bird's syrinx. The extra control is provided by two things. First, the shape of the membranes, and hence the type of noise they create, is finely regulated by a series of muscles and soft bony rings on the outside of the syrinx. Second, instead of having just one vibrating membrane, like our blade of grass, birds have two, one on each side of the syrinx.

As part of his study of birdsong in the 1770s, Daines Barrington persuaded pioneer surgeon John Hunter to cut up a selection of songbirds for him in an effort to establish exactly how they made their various calls.[5]

I procured a cock nightingale, a cock and hen blackbird, a cock and hen rook, a cock linnet, and also a cock and hen chaffinch, which that very eminent anatomist, Mr Hunter, F.R.S. was so obliging as to dissect for me, and begged, that he would particularly attend to the state of the organs in the different birds, which might be supposed to contribute to singing. . . . Mr Hunter found the muscles of the larynx to be stronger in the Nightingale . . . and in all those instances where he dissected both cock and hen that the same muscles were stronger in the cock.

So the syrinx, it seems, is better developed in males than females, and in birds with more elaborate songs. Later researchers found that the variation in syrinx design among different species was so great that it could have been used to establish the evolutionary relationships between different birds. At the same time this variation in design has completely frustrated those trying to understand how the syrinx actually works. Researchers summed up the syrinx like this:[6] It 'should be a morphologist's dream. It is relatively isolated, contains a variety of muscles, and is replete with levers and potential oscillators. Yet that dream has . . . proved to be a nightmare.'

Most of us, armed only with our more modest larynx, can readily make noises similar to many birds (and of course some birds, including the canary, can make noises like the human voice), but what requires immense skill and training is stringing these noises together to produce a 'song'.

No one knows how the avian brain controls song, but we do know that across different bird species the quality of song varies with the size and complexity of both the syrinx and the brain. As birds begin to sing in spring, the part of their brain that controls and remembers song, the higher vocal centre, increases in volume (and decreases once the breeding season is over). The male canary's higher vocal centre is larger than the female's, which is not unexpected given that males sing and females do not. The same region of the brain is relatively larger in those species that utter more sophisticated songs. But females still need enough neural circuitry to make sense of the male's song. We also know that female canaries have the potential to sing and, given a shot of testosterone, this is exactly what they do. Scurrilous traders in the past dosed female canaries with testosterone and passed them off as males because they were worth more. Remarkably, one of the things testosterone does to a female canary is to stimulate that region of the brain concerned with song to enlarge by growing new cells[7] – something that was once considered impossible.

From a female canary's point of view the key elements in a male's song are those bits that tell her about his quality as an individual – the sexy syllables. If we look at a sonogram of these sexy syllables they look very special indeed. They are two-note trills repeated very rapidly – up to seventeen times a second – and rather like Tuvan throat singing (only better), or a musician playing an Etruscan flute or Picasso simultaneously painting a separate image with each hand. The canary's two-voice syllables are made up of a low- and a high-frequency sound (Figure 2). Now recall the dual nature of the bird's syrinx: the high-frequency sound comes from the right-hand side and the lower sound from the left. The simultaneous production of these sounds by the male canary requires extraordinarily sophisticated muscle coordination and it is precisely because these phrases are so difficult to produce that they accurately reveal a male's quality. Only genuinely top performers can utter such intricate and intriguing sounds – they simply cannot be faked. This is the essence of a successful sexual display – it must honestly reflect a male's quality. If all individuals could produce double-note trills they would be useless cues for females to distinguish great from mediocre performers.

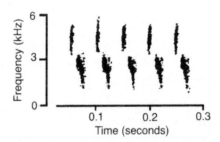

FIGURE 2 *A sonogram of the canary's sexy syllables, here repeated just five times in a fraction of a second.*

Acquired Choristers

Karl Reich, of course, knew nothing of this research, which lay years in the future. He knew only that he succeeded where all before him failed. As they finished their cigars, Reich explained to Duncker that while others in the past had produced nightingale-canaries, he was the only person to *breed* an entire race of canaries whose exquisite song was so obviously inherited. The evidence, Reich said, was indisputable – his birds' songs got better and better with each new generation and, what's more, they continued to improve without hearing the merest strophe from a real night-ingale. Impressed by what he had seen and heard, Duncker knew that Reich's explanation, grounded as it was in a Lamarckian view of evolution, must be wrong.

Jean Baptiste Pierre Antoine de Monet, Chevalier de Lamarck, as he was formally known, believed that useful changes acquired during an individual's lifetime become incorporated into what we now call the genome and could be passed on to succeeding generations, a mechanism referred to as the 'inheritance of acquired characteristics'. The giraffe was the perfect example, their long necks the result of generation after generation stretching up to pluck leaves at the tops of trees. Lamarck's evolutionary mechanism was an optimistic one with deep intuitive appeal, for it seems only right that individuals should be rewarded for a lifetime of endeavour by having their efforts incorporated into their genetic heritage.

Breathless, Reich finished his soliloquy. His birds were his pride and joy, and his achievement had made him something of a celebrity across Europe. Outwardly enthusiastic and genuinely impressed by Reich's birds, Duncker was privately sceptical that their song was genetically inherited. At the time, however, Duncker could not come up with a convincing explanation for Reich's startling results.

Lamarckism was the most tenacious form of 'soft inheritance' and died only slowly. After the publication of the *Origin of Species* many people believed in evolution, but not everyone was convinced that Darwin's notion of natural selection was the only mechanism by which species changed over time. There were two problems. Natural selection seemed sensible, but it lacked a coherent theory of inheritance and it was unclear whether it could ever be enough to bring about major changes such as the origin of new species. There was no doubt that a bird's wing was a wonderful adaptation for flight, but many found it hard to see how it could have been created bit by bit by natural selection. What use would half a wing ever have been? For most people, including many biologists, natural selection on its own was not enough. Something extra was needed and the idea of 'directed mutations', that is, changes in tune with what an individual 'wanted', seemed to fit the bill. Evolution with a purpose was what many felt a benevolent God would have wanted. In other words, the hereditary process (whatever it was) produced adaptations automatically and allowed those characteristics acquired during an individual's lifetime to be passed on to their offspring. The inheritance of acquired characteristics is now referred to as Lamarckism, even though Lamarck didn't invent it (the idea went back to Plato), nor was it Lamarck's main process of evolution. Nonetheless, the name stuck and Lamarckism was natural selection's main opponent during Darwin's time.

Later, in the 1880s, the brilliant German biologist August Weismann pointed out the utter fallacy of Lamarckism. His great insight was recognising that the germ plasm (the reproductive cells) was separate and distinct from the soma (the body) and nothing that happened to an individual's body during its lifetime could be communicated to the reproductive cells. He was therefore one of the first to appreciate the difference between an animal's appearance – its phenotype – and its genetic constitution – its genotype. Nonetheless, despite Weismann's findings and Mendel's convincing hereditary mechanism twenty years later, many biologists continued to believe in the inheritance of acquired characteristics. There was no rational scientific reason for this, they were simply reluctant to relinquish the comforting idea that evolution was progressive, just and driven by self-improvement. Not until about 1930 did Lamarckism finally start to fade away among mainstream biologists.

That Reich believed in a Lamarckian process of heredity was hardly surprising. He was a shopkeeper, not a scholar, and like many others before and since he was beguiled by the sense of natural justice inherent in a Lamarckian heredity. Almost all his bird-keeping friends believed in it and the newspapers in the 1920s were full of stories of some ingenious Lamarckian experiments conducted by the Viennese scientist Paul Kammerer.

Just before the First World War Kammerer claimed to have demonstrated the inheritance of acquired characteristics in several different animals – his best-known study being on the midwife toad. Most species of toads breed in water and to help them hang on to females the males develop a horny thumb pad. Because midwife toads breed on dry land, however, they have no need of thumb pads. What Kammerer did in his experiment was to force midwife toads to breed in water, and they developed thumb pads! Despite his apparent success, Kammerer's experiments failed to attract the attention he felt they deserved. So after the war he reinvented himself and set off on a pro-Lamarckian publicity tour, denouncing followers of Gregor Mendel as 'slaves to the past' and touting Lamarckists like himself as 'captains of the future'. He became a media star and persuaded Reich and many others to accept Lamarckian inheritance.

Kammerer remained in the spotlight only briefly. In 1926 the American scientist G. K. Noble asked to examine Kammerer's voucher specimen – the single lecture-tour toad Kammerer had preserved for posterity – only to discover that the dark thumb pads were fakes created with injections of black ink. Devastated, Kammerer walked into the Alps and shot himself. Kammerer's supporters proclaimed his innocence – someone else, possibly his assistant, had made the injections; a helping hand to generate the results Kammerer so desperately hoped for. The other rumour was that the Nazis had tampered with the toads to discredit the Jewish Kammerer because Lamarckism was antithetical to their beliefs.[8]

Reich's beautiful songsters continued to tax Duncker, but he was absolutely convinced that the answer did not lie in Lamarckism. The puzzle forced Duncker to ponder how canaries and other small birds actually acquired their song. How much of it was genetic and how much was learned? For over two centuries there had been a tremendous incentive to answer this question – from a practical standpoint, if not a scientific one.

Hervieux's handbook, published in 1705, provided clear and sensible instructions for training canaries to sing.[9] He recognised that the young birds had what biologists and psychologists now call a 'sensitive period' in which they learned their song, and that to perfect their performance they must have no distractions at this critical time. Hervieux condemned the then common practice of putting young canaries in a dark box – from which few emerged alive – to help them learn: 'If you desire to succeed better in that Point you must observe this Method.' A few days after they are capable of feeding themselves, young birds should be placed individually in a cage covered by transparent linen in a room distant from all other birds whatsoever, so that he may never hear any of their Wild Notes, and then play to him upon a little Flageolet' . . . After a fortnight the linen cover was to be replaced by one of 'green or red Serge . . . till he perfectly learns what you teach him'. Hervieux recommended feeding the birds only at night by candlelight, the idea being that if kept without any visual distractions the young bird would learn more rapidly. 'As for the Tunes, he must be taught only one fine Prelude and a choice Air; when they are taught more, they are apt to confound them.' Hervieux urges patience – 'for without it nothing can be done' – and finally, 'I have thought fit to prick down the Prelude and Air I have spoken of' (see Figure 3).

Seventy years later Daines Barrington, in his 'scientific' study of birdsong, confirmed Hervieux's assertions by rearing young linnets with different species as song tutors and showing that they learned only their tutor's songs. He found that young birds continued to 'record', that is, learn their song, for ten or eleven months. 'These facts . . . seem to prove very decisively, that birds have not any innate ideas of the notes which are supposed to be peculiar to each species.'[10]

But Barrington was not wholly correct. Research conducted in the 1950s – paradoxically much of it on chaffinches rather than canaries – revealed that most songbirds do have an inbuilt concept of what they should learn. When young chaffinches were hand-reared and brought up without ever hearing a chaffinch song, they spontaneously began to sing something that most birdwatchers would recognise as a chaffinch song, but by comparison with wild birds they uttered only a very simple tune.

Isolation became a central part of song-learning experiments. The birds used in these studies were called 'Kaspar Hausers', after a dumb and helpless teenage foundling discovered in Nuremberg market place in May 1828. Allegedly the illegitimate son of the Grand Duke Charles of Baden, Kaspar Hauser was kidnapped as a baby and reared in isolation by peasants. After being rescued and learning to speak, it transpired that he had been kept in a hole and fed by a man who, after teaching him to stand and walk, abandoned him in the market place.[11]

The chaffinch is slightly unusual compared with other finches because, as Pernau, Barrington and Hervieux knew, canaries could easily be persuaded to learn the song of another species, whereas a chaffinch could not. This does not mean that the canary lacks an innate song template, only that its template is more flexible than that of the chaffinch.

FIGURE 3 *An air that might be taught to canaries, from Hervieux's book (1705) on the canary.*

Flageolets and Serinettes

Hervieux trained his canaries with a flageolet – a small flute made of ivory or wood – its pitch precisely suited to match the canary's vocal range. The term 'recorder' seems to be intimately associated with the training of singing birds. We now think of the recorder as a child's instrument, but to 'record' meant to 'remember' and the recorder's original purpose was to help a canary or other songbird remember its tune. A small volume published in London in the early 1700s, *The Bird Fancyer's Delight*, thought to have been written by the

Staffordshire ornithologist John Hamersley,[12] exploited the contemporary fashion for training songbirds. It contained instructions similar to Hervieux's for the canary, and no less than forty-three different tunes: eleven each for the bullfinch and canary, six for the linnet and the rest for various others. Some of the tunes, such as Handel's *Rinaldo* and 'The Happy Clown' from *The Beggar's Opera*, were already popular, but many appear to have been especially written by the next-to-anonymous Mr Hill – a flageolet-playing friend of Samuel Pepys, who was himself a flageolet enthusiast.[13]

The tunes in the *Bird Fancyer's Delight* are generally rather jolly and often incorporate trills and other birdlike sounds easily fingered on the flageolet. The number of tunes allotted to each species probably reflects the facility with which different birds responded. The bullfinch and the canary had most tunes and were certainly the most popular singing birds. The woodlark has four tunes, the skylark and starling three and the nightingale just two. The sparrow, throstle (song thrush) and India nightingale (possibly the mynah), were given just one each.

Charming as it is, Hamersley's little book was probably little more than a scam. Two hundred and fifty years later, when song-learning had become an important topic of scientific research, experts questioned whether birds could really master the rather complicated tunes Hamersley set out for them. It seems that Hamersley's book was aimed at two types of people – those interested in practising the flageolet but with no interest in birds and gullible bird keepers who were too inexperienced to know that the tunes were too difficult for their pets.[14] Scam or not, the *Bird Fancyer's Delight* stayed in print for over a hundred years and remains popular among recorder players today.

Teaching canaries and other birds to sing particular tunes grew so popular that it was perhaps inevitable that someone would eventually design a small machine to save them the effort of puffing on pipes or flageolets. The early eighteenth century was a period of great inventiveness in Europe: the flying shuttle, the seed drill and more efficient steam engines revolutionised weaving, agriculture and industry. In France around 1730 a Monsieur Bennard designed and constructed the first serinette – a miniature barrel organ, with bellows and ten tin pipes – which, at the turn of a handle, produced a high-pitched tune suitable for a small bird to learn. The serinette was the first way of recording music and for those who could afford one it must have been rather like today's Macintosh iPod – the most up-to-date of musical gizmos and an important status symbol. George III, who loved gadgets, had one that played several different tunes, but to little effect for in a fit of rage he killed his birds when they failed to learn as quickly as he expected. Serinettes feature in several well-known contemporary works of art, including William Hogarth's 1742 painting of the well-to-do Graham children in which the serinette is operated by the boy, but the 'pupil' – a goldfinch – looks more interested in avoiding the children's cat than learning a tune. In Jean Baptiste Chardin's *La Serinette* (1751) the artist's second wife (acting as model) is seated in another opulent interior with the bird organ on her lap and a canary x goldfinch hybrid in a nearby cage.

Serinettes came in various models, generating sounds of different pitch, and you chose one appropriate for the bird you wanted to train. The commonest type was the canary serinette, but there were versions for the blackbird and bullfinch, and even one for the parrot. Despite their inventor's ingenuity, it was the neural abilities of the birds themselves – at least under this form of tutorage – that provided the ultimate limit on what could be achieved. The way they sang often left much to be desired. J. S. Bach's son Carl Philipp Emanuel once exhorted his pupils to 'play from the soul, not like a trained bird!'.[15] With these limitations it is surprising that the serinette fad lasted as long as it did. But as the serinette sank into obscurity in the mid-nineteenth century it was replaced by a new fashion: bullfinches trained to mimic a human whistle.

Whistling Gimpels

Daines Barrington deliberately excluded the bullfinch from his study because it couldn't sing without 'instruction'. Its natural song was considered about as musical as a squeaky wheelbarrow. Despite this lack of innate vocal talent, in captivity – as Hamersley makes clear – it had been known since the Middle Ages that a young bullfinch could learn 'almost any tune that is taught by pipe or whistle if not too long'. In addition, the male bullfinch is among the most gorgeous of European birds, with a luscious pink breast, black cap and slate-grey back, making it instantly appealing despite its reputation for being somewhat delicate as a cage bird. It was easily procured and earned its German nickname '*Gimpel*', or fool, because it responded so predictably to a decoy bird and was easy to capture. When kept as a pair, bullfinches show an endearing fondness for each other that reflects their lifelong pair bonds in the wild.[16]

Starting in about 1850, training bullfinches to whistle folk tunes became a specialist occupation, and an important source of income for farmers and craftsmen in the Vogelsberg area of Germany. Piping bullfinches became incredibly popular and large numbers of them were sold (expensively) not only in Germany but also in Holland and England. When Darwin went to London to visit the German bird artist Joseph Wolf in 1870, he was greeted by Wolf's pet bullfinch which immediately started to search for crumbs in Darwin's beard. Like other finches, young male bullfinches learn their natural song from their fathers. To produce a piping bullfinch the young birds were taken from the nest and reared in isolation. They saw only their keeper, who whistled two German folk tunes to them – in the same order and same key – several times each day for about five months. After this the young bullfinch retained these tunes for life – if it was a male. Unfortunately for the trainers, the sexes of young bullfinches are indistinguishable and about half their effort was wasted on females

who very rarely learned to whistle. Even the males sometimes failed
to learn the tunes.[17]

William Cowper's poem 'On the Death of Mrs Throckmorton's
Bullfinch' (1788) paid tribute to the bird's learning abilities:

> And though by nature mute
> Or only with a whistle blessed,
> Well-taught he all the sounds expressed
> Of flageolet or flute

That a bullfinch can be taught to whistle a German folk song and a
canary trained to repeat the song of a nightingale might imply that a
bird's brain is little more than a biological tape recorder. But a remarkable study by the German biologist Jürgen Nicolai in the 1950s shows
what extraordinarily sophisticated mental processes are involved in
birds acquiring their song.

The bullfinch is an enigma. For a bird with no natural song to
speak of, its ability to learn tunes is absolutely amazing.[18] Fascinated
by this paradox, Nicolai decided to explore the full extent of the
piping bullfinch's hidden talents, studying wild birds in a local ceme-
tery and captive ones in his own home. As an old man, Nicolai told
me how his studies started when he bought a tame bullfinch, which
had been advertised in the local newspaper. He never looked back:
fascinated by this bird's ability to sing, Nicolai began a comprehensive study involving fifteen folk-song-singing male bullfinches and
their male trainers.

The first thing he found was that the young birds are extraordinarily motivated to perfect their songs, repeating them again and again,
and just like a child, starting again from the beginning whenever they
make a mistake. Second, when Nicolai made tape recordings and
sound spectrograms of bullfinches and their trainers, he found that
the birds always transposed the tune to a higher pitch, from F to F#,
for example, or from G to G#. Astonishingly, the birds also seemed

to have an inherent appreciation of exactly how a tune should go and usually performed it better than their trainer. Even if the air flow of the whistling trainer was slightly irregular, the birds always uttered a smooth and completely uninterrupted sequence of notes. The birds also had an uncanny feel for their tunes. If the trainer whistled the next phrase of a tune ahead of the singing bird, the bird would stop, wait and then continue where the man left off, completing the tune. Bullfinches could also recognise tunes as separate musical entities. Nicolai's inspection of the spectrograms showed that trainers generally separated two tunes by a short pause, but no longer than the pause between different phrases within a tune. Yet the birds always sang the two tunes separately.[19]

There's a final twist to the bullfinch's tale. What gives us a preconceived idea of what we are looking for when searching for a partner? Among birds the notion of what is attractive – or sexual imprinting, as it's called – is put in place in the first few months after leaving the nest. Young bullfinches were traditionally taken before fledging and reared and trained by men, and as a result became sexually imprinted on men – meaning that as adults they directed their sexual approaches to men, assuming them to be what they should establish a long-term relationship with. Trainers sold their young birds at the end of their education when they were four or five months old. Within just a few days they had transferred their affection to their new *male* owner and soon greeted him exactly as they would a sexual partner, with an endearing display of fluffed body feathers and twisted tail, piping their learnt tunes. Women, however, were rejected and birds reacted to them as though they were a completely different species. A bullfinch 'which a lady (north of the Tay), bought from a German bird-dealer, because of the excellence of its song, was no sooner in her possession, than it became entirely mute, and, though apparently in perfect health, neither voice nor instrument could induce it to sing'.[20] One wonders how many other disappointed lady customers the German dealers left in their wake.

Bullfinches can clearly distinguish between their male and female owners but these observations do not mean that they possess any innate preference for men over women. I suspect that had young bullfinches been trained by women, they would have been sexually imprinted on them as adults. The experiment, however, has never been done and may never be done. The tradition of training bullfinches died out in the 1970s, when the last trainers passed away.

Managing to Sing

For German song canaries, including Reich's birds, singing is a product of both the right genes and the appropriate learning opportunities. But these qualities alone are insufficient to produce a champion. A third component is needed – one that Reich, like all Harzer breeders, knew all about. Young roller canaries require very careful management if they are to sing properly. The trick is to train them to sing in response to light and this is why competition birds are kept in small wooden cages with a pair of doors that open over the wire front. The birds learn to find their food and water in the dark, and the doors are opened for just ten minutes in the morning, thirty minutes around midday and another ten minutes in the evening. In this way, as Reich demonstrated to Duncker, the birds learn to sing when the cage doors are opened, although there is a bit more to it than a simple conditioned reflex.

The amount of light the bird experiences regulates, via its hormones, the amount and quality of its song. One roller breeder told me that if he were to keep his competition birds in an ordinary cage or aviary they would be over-stimulated by the light and sing too much – what he called 'screaming'. Roller experts also use diet to regulate their birds' performance. Most canaries are fond of minced hard-boiled egg and many breeders feed this to their birds, especially during the breeding season. But for a roller breeder, a diet containing egg is a disaster since, for some reason, it causes the birds to sing flat. You can also enhance the singing ability of an underachieving bird by feeding it tonic seed and subdue an overeager singer by feeding it poppy seed (maw).

Male displays like plumes or chorister voices would be pointless if females did not possess the mental ability to assess and grade them. In much the same way those who judge the products of artificial selection need to be up to the job. It required relatively modest neural circuitry for men to grade big-buttocked cattle, but judging roller canaries takes immense skill. One needs a very good ear and years of training to become a roller canary judge. I have listened to roller canaries going through their 'tours', as they are called, while a top judge explained each component to me. I was embarrassed to find the 'glucks', 'bell rolls' and 'flutes' almost impossible to distinguish.

Amazing Ascendancy

After a year of friendship, Duncker and Reich decided in 1922 to go public and present Reich's best birds to the newly formed national committee for judging canary song, the Deutsche Einheitsskala, in Kassel. Reich's birds were unanimously acclaimed by the judges who had never heard anything like them before. Backed by this official approval Duncker wasted no time in publicising his own ingenious interpretation of Reich's achievement. He wrote articles for several different magazines, spanning the full range of bird enthusiasts, from the amateur canary keepers to evolutionary biologists. In so doing Duncker not only thrust Reich's birds into the public eye, he also placed himself in the scientific spotlight. This was no whim, but rather a carefully orchestrated strategy. In a paper he published in the premier German bird journal, *Journal für Ornithologie*, Duncker described the extraordinary nature of Reich's self-sustaining strain of birds. He explained that the birds no longer needed a nightingale tutor and then restated Reich's view that through careful breeding his canaries had been permanently modified by hearing nightingale song. How could this have happened? Duncker then played his trump card. Rather than selecting for the nightingale song itself, which is what Reich believed he had done, in fact he had unconsciously selected birds for their genetic ability to learn the song from their nightingale-canary tutors. This was a brilliant and original explanation. As Duncker reminded his readers, there was no mechanism by which one bird's genes could be modified by the sound of another. Reich's Lamarckian style of inheritance simply did not work. Duncker's idea, on the other hand, was rooted firmly in a straightforward Darwinian-Mendelian process.[21]

Confident that he was correct, Duncker spelled out exactly how his idea could be tested: one could take some of Reich's birds and rear them either with ordinary canaries or in a situation where they could hear no other birds. If birds reared in either of these ways subsequently sang a nightingale song then Reich's Lamarckian view would be right, but if they sang anything else, Duncker's hypothesis would be confirmed.

Duncker was so far ahead of his time that this part of his collaboration with Reich has been completely overlooked by all subsequent birdsong scholars. Twenty years later when song learning in birds caught the attention of the new field of animal behaviour, a single tangential, slightly deprecating reference to Reich's work appeared in Julian Huxley's *Modern Synthesis* (one of the most influential biology textbooks of the 1940s).[22] In a footnote on page 305, Huxley says that Ernst Mayr – student of Erwin Stresemann and later the grand old man of evolutionary biology – had told him about a study in which canaries had been taught to sing using recordings of nightingale song 'carried out by a fancier named Reich, but complete proof was not supplied'. So near and yet so far! Even in the 1950s and 1960s the great pioneers of song learning, Bill Thorpe and Peter Marler, failed to notice Duncker's ingenious interpretation of Reich's wonderful experiment. To this day, many professional ornithologists continue to look down on amateur bird keepers, partly because bird keeping is déclassé and partly out of intellectual snobbery.

Fortunately, other aspects of Duncker's and Reich's collaboration were more successful in breaching the barricades of science.

5

The Variegation Enigma

They who couple a Grey Cock and Grey Hen, which are both common, can expect none but a Grey Breed. But when different kinds are mix'd it falls out better, for Nature often delights in producing finer and more beautiful Birds than was expected.

HERVIEUX, A New Treatise of Canary Birds (1718)

Reich's bird room soon became a second home for Hans Duncker, and most evenings after school he would join Reich for a chat and a cigar. They had resolved the nightingale song issue and Duncker was looking for a new challenge. One evening as Duncker peered through his tortoiseshell-rimmed spectacles into a cage full of canaries he asked Reich about the appearance of his canaries. Why were some yellow, some a uniform dark-green and yet others variegated – a mixture of the two colours? Reich replied that most canaries were like this and colour was immaterial for those who were interested only in their song. He then added as an afterthought that in another branch of the fancy there were breeders who specialised in coloured canaries, like the new whites, or a washed-out brown variety called cinnamon. The English, Reich said, had a breed whose plumage resembled a reptile's scales; the lizard canary, and there was even an orange-feathered bird, the Norwich – named after the city that made it famous. For good measure Reich mentioned that almost all the English 'type' canaries – those bred primarily for their body form – like the Norwich,

Yorkshire and Lancashire Coppy also existed in green, yellow and variegated versions. Fascinated, Duncker thought about this for a while and then asked whether anyone knew how the different colours were inherited. Reich told him that as far as he knew it was pot luck and no one had ever studied this in detail. Duncker's eyes lit up and he proposed there and then that they do the experiments themselves. Reich was game and they started to plan the pairings that would establish whether the colour of the canary's plumage was inherited in a predictable Mendelian manner.

It was well known among canary enthusiasts that wild canaries were green and that the yellow colour of the domesticated birds was a mutation. Duncker wanted first to establish whether green plumage was dominant over yellow, just as roundness was dominant over wrinkledness in Mendel's peas. Before they started, he decided to check whether anyone else had previously investigated the inheritance of canary colours. From his conversations with Reich he knew that canary fanciers had written on the subject, but he was surprised to discover that several scientists had also studied canary colours. There were two relevant accounts,[1] both published in 1908. The first was an article by an English woman, referred to as 'Miss Durham', on the inheritance of cinnamon plumage. The other article was by a well-known American biologist, Charles Davenport, who had investigated the inheritance of both canary crests and colours. The two papers could not have been more different. Miss Durham's was concise and precise, a model of clarity, and her results were reliable because they were based on hundreds of canary offspring. The other paper was a muddle from start to finish, and Duncker found it difficult to see how Davenport could so confidently conclude that both colour and crests obeyed Mendelian rules. To make matters worse, Davenport had used very few birds in his research, casting further doubt on his conclusions. On the plus side, Davenport's paper did include a fascinating account of the canary's early history and domestication. Duncker later found that he was not alone in questioning

Davenport's conclusions, and that this particular study had engendered a bitter dispute among fanciers and geneticists alike.

What was absolutely clear from both accounts was that the issue of colour was a long way from being resolved, so Duncker set about designing his own experiment to establish the genetic basis of colour once and for all. At the beginning of 1923, Duncker and Reich paired together different combinations of the three basic colour forms: yellow, green and variegated birds – seventy pairs in total – which, by July, under Reich's superb management had reared a staggering total of 517 youngsters. Some of the results were clear-cut: two pure-bred yellow parents generally produced yellow offspring, and green parents always hatched green babies. But it was the other pairings that yielded the most intriguing results. Pairs comprising one green and one yellow bird invariably produced variegated offspring and variegated birds paired together tended to produce offspring which varied enormously – from pure yellow to almost completely green, with everything else in between. Duncker was intrigued to see that the patches of green plumage on these variegated birds were not distributed at random over the birds' bodies and even in birds that were predominantly yellow, patches of dark plumage persisted rather predictably around the eye and on the wings. It was these plumage patterns that led him to the realisation that the canary's colours were controlled by several different factors (what we now call genes) – each responsible for a different region. Despite this unexpected complexity, Duncker was able to conclude, with a sense of satisfaction, that canary colours did indeed follow Mendelian rules.[2]

Between bird-room conversations about their breeding results there seems little doubt that Reich and Duncker discussed the increasingly desperate state of the economy. As inflation spiralled out of control and they saw their hard-earned savings become worthless, both men must have felt distinctly gloomy. The inflation rate was so high that they joked about meals increasing in value as they were being eaten. But this was no joking matter. Since the end of World

War I the harsh peace terms imposed by the Allies had rendered the German economy extremely fragile. It was made worse by the Allies' reparation demands that Germany attempted to fulfil by massive borrowing, which in turn gave rise to the inflation. The middle classes – people like Reich and Duncker – were especially hard hit and later made them a rich recruiting ground for the National Socialists. The Weimar Republic, formed in 1920, was a government born of defeat and one that nurtured political unrest. Then, in November 1923, as Duncker prepared the results of the variegation experiments for publication, the unrest exploded and sixteen protesters were shot dead on the streets of Munich as Adolf Hitler attempted to seize power. Ever since 1921, when he became chairman of the National Socialist Party, with its programme of 'anti-Semitism, simplistic economic theory and pseudo-socialist rhetoric', Hitler had been agitating for reform. His November coup, however, was swiftly foiled by the police and he ended up in prison, where he started to document his struggle for existence, *Mein Kampf*, and for a while, at least, things returned to normal. In fact, they started to improve, for in January 1924 the Allies reduced the reparation payments, marking what was for many Germans the real end of the war. The period of economic stability that followed saw a cultural revival during which artists like Paul Klee and Wassily Kandinsky at the Bauhaus flourished, and Hans Duncker was at his most productive.[3]

Duncker went back to Göttingen and to the library. Little had changed in this medieval town since his student days, but there was one thing he was keen to see – the Gänseliesl or goose-girl – who stood in the centre of town in front of the town hall. The most popular girl in Göttingen, she reputedly received a kiss from every student (most of them were men) on the day they graduated. Hans smiled at the idea as he walked past her statue on his way to the library. There he re-read Davenport's 1908 account of the canary's early history and his confident assertion that the canary's transition from green to yellow had occurred in just a few generations. The idea that this change, which was thought to have happened around 1700, was a swift one was consistent with the view held by a group of scientists, including Davenport, who referred to themselves as 'Mendelians'. They held that most domestic forms of animals and plants arose more or less spontaneously as 'sports', which were then rapidly fixed by artificial selection. But Davenport's claim regarding the canary's rapid change flew in the face of Duncker's and Reich's recent results. Duncker knew that with several separate genes involved, the green-to-yellow transition could not possibly have been rapid and, in fact, must have been painfully slow. Had the canary's yellow plumage been controlled by a single gene – a single mutation – the transition certainly could have taken place within only a few generations, equivalent to having to get just one number in a lottery. But because it was clear from Duncker's experiments that the canary's plumage colour was polygenic and controlled by several separate genes, the change from green to pure yellow would have taken many generations. Capturing a polygenic trait is more like trying to get a specific set of numbers in a lottery and much harder than getting a single one, and hence requires many more attempts.

Royal Distractions

When Karl Reich's nightingale-canaries were rated 'outstanding' by the Kassel judges in 1922 the media dubbed Reich as the 'new Hervieux'. This was praise indeed for Joseph-Charles Chastanier Hervieux de Chanteloup was *the* canary pioneer. His book, *A New Treatise on Canaries* (1705), was the first ever monograph on canaries and Hervieux's ability to train canaries to sing was unsurpassed. In short, Hervieux was the ultimate canary connoisseur and when Hans Duncker started to track down some of the old accounts of canary history, it was to his wonderful little book he turned first. The successive editions of Hervieux's *New Treatise* provided an unrivalled set of snapshots of canary culture through the eighteenth century.[4]

Hervieux was chief canary wrangler to the Princesse de Condé, also known as Madame la Princesse. She was the long-suffering wife of Henri-Jules, the Prince de Condé, and they lived in the palace at Chantilly, just outside Paris. Ugly, anorexic and alcoholic, the prince was a sad case. His father, the Grand Condé, was an outstanding soldier with high hopes that his son would follow in his footsteps, but Henri-Jules's attempts at soldiering failed miserably. The prince married the fifteen-year-old Anne of Bavaria in 1663, making her the Princesse de Condé, but like so many royal marriages this was a political and tedious union, whose main purpose was to inject some fresh blood into an increasingly inbred aristocracy. The strategy worked and the marriage proved to be a productive one. Apart from siring heirs – ten in total – Henri-Jules passed his time dabbling in the arts and with science, but he excelled only in reckless and extravagant entertaining. As he aged he became more and more erratic, terrorising his wife and children with malicious practical jokes. The princess took refuge in Chantilly's vast menagerie established many years earlier by her father-in-law. The canaries were her favourites and some time in the 1690s she employed Hervieux to look after them. The princess was by then middle-aged, and the production of ten children and a lifetime of marital torment had taken their toll, but she was warm-hearted and generous to those she liked, and Hervieux quickly became a favourite. Young Hervieux and the canaries proved to be a perfect distraction. She revelled in his rearing, nursing and training of her canaries, and for him royal patronage was more than he could ever have hoped for.

Hervieux was a sharp operator: in addition to his sure touch with birds, he knew how to please his royal mistress. He could train canaries to sing particular airs and had invented a flageolet especially for this purpose. He also orchestrated spectacular canary singing concerts for the princess and her children. Encouraged by the princess,

Hervieux was just twenty-two when his book was published in 1705. The timing was impeccable. Canaries had become increasingly popular, particularly among the female aristocracy, and a reliable account of their care, breeding and training was much needed. The book was extraordinarily popular and went through no less than ten French and several foreign editions, and remained in print for over a century. Hervieux knew his canaries and wrote beautifully, and much of his information is as applicable today as it was 300 years ago. Despite his success, however, Hervieux himself remains something of a mystery. No portrait exists, but legend has it that even at the relatively youthful age of twenty-five he looked like a bird, with a huge hooked nose and a *petite complection*. It was said that in later life he did not sleep in a bed but instead roosted on an armchair. His official title was 'governor of canaries' and reading the dedication to his *Princesse*, one might be forgiven for imagining that he was in love with her. He might well have been, even though she was thirty-five years his senior, but Hervieux also knew on which side his bread was buttered and at the time such sycophantic dedicatories were de rigueur.[5]

I take the liberty of offering your Highness this small work, which already belongs to you, because it is only on account of the canaries belonging to your Highness that I have undertaken it. The title of governor of the canaries, with which she was kind enough to honour me herself, gave birth to the idea of making these comments: and I thought that to deserve the title it was my duty to work upon the conservation of these charming little birds which sometimes entertain and relax the spirit of your Highness. In order to succeed in this, I have dedicated myself with particular care to gathering everything that might be necessary, as much for their rearing, as for their preservation [keeping them alive]. The innocent pleasure which your Highness can take, is not unworthy of the high rank which she holds, nor of the improvement of the mind which makes her shine everywhere, since St Jean, that great saint, and preoccupied as he was, continually meditating upon celestial things, but was still able to relax with his Partridge. Since, madame, you emulate the saints in their virtues, you can also emulate them in their spiritual pleasures, which they bring back to God, admiring in the smallest creature the infinite wisdom of the creator. I would consider myself very fortunate if your Highness would be so kind as to grant this book the honour of your protection from which alone it can draw all of its merit; and allow me to say with very profound respect and deep gratitude of your serene highness, very humble, very obedient and very indebted servant, J. C. Hervieux.

Hervieux's book marked a turning point in bird keeping. Among other things, it inspired a new profession, the 'siffleur d'oiseaux', the bird whistler; the peripatetic canary teacher, the first personal trainer. Three times a week the siffleur turned up at the homes of bored and wealthy women to train their canaries to sing. A variety of instruments was Used to create these airs; a flageolet, the German flute and a small water-filled whistle, which produced a bubbling or warbling sound. You can still buy ceramic versions of these in Portuguese gift shops where they are now sold as children's toys. As well as describing how to train a young canary to sing, Hervieux's book provided details of how to breed canaries, which the French had to this point largely imported from Germany. More significant for Duncker, Hervieux listed no fewer than twenty-eight varieties of canary existing in the 1700s, providing a tantalising glimpse of the domestication process. Much has since been made of Hervieux's extended list and it now seems that most of his varieties were simply colour variations rather than separate strains as we would define them today. Baron von Pernau, writing at about the same time as Hervieux, but from Germany, had a rather less expansive list, recognising only five main varieties: dark yellow, pallid yellow, partly black (that is, variegated), white and mealy – the colour of bread.[6]

Shipwrecked Plagiarists

Where did the canaries that kept Hervieux in business, distracted Madame la Princesse from her lunatic husband and made Reich famous come from? From the Canary Islands, of course, albeit by a circuitous route. Almost every book ever published on canaries starts with the story of how in the sixteenth century a ship bound for Leghorn (Livorno) with a cargo of wild canaries from the Canary Islands was wrecked off the island of Elba, off the north-west coast of Italy. Miraculously, some of the birds escaped the sinking ship and flew to Elba. As was typical at that time, the exported canaries were all males and in order to propagate themselves in their new home – so the story goes – they were forced to breed with some of Elba's resident female serins. Their distinctive offspring in due course formed the basis for many of the canaries that would later be exported across Europe. The shipwreck story is usually attributed to Giovanni Pietro Olina, whose lavish volume on bird catching and keeping, *L'Uccelliera*, published in Rome in 1622, tells the tale, but it actually appeared first in an obscure book published by another Italian, Antonio Valli da Todi, twenty-one years earlier.[7] Here is Todi's original version of the story: 'On the Island of Elba one can also find cross-bred Canaries descended from true ones, for the following reason: a ship coming from Canaria to these parts was wrecked over the rocks of this Island and, carrying many of these birds, they arrived on the said island where they are [now] found; they are as large as a siskin, but much yellower on the chin than the true canary, and have black feet, and such is the cross-bred male.'

Subsequent writers embellished the shipwreck legend, a process analogous to the errors that occur in the copying and re-copying of DNA from generation to generation. Here is a version from an anonymous canary book published in 1873:[8]

A bark, laden with dried fruits, spices, and canaries, came to grief on the coast of that pleasant Mediterranean island [of Elba]; when the birds escaping, found the climate suitable to their constitution, and betook themselves to its orchards and other shady and leafy retreats, where they claimed a republic, dreamt of liberty, and revelled in the idea of never-ending enjoyment. But alas! Poor birds, their anticipations and visions of freedom were shortlived; for the attention of the peasants being attracted by their beauty and their sweet notes, traps and nets were set for the little waifs and strays thus thrown on their island; and ere long the captives were dispersed over the continent of Europe.

In this account the notion of the birds being cross-bred or having hybridised with the local finches has disappeared altogether. Other versions assert that descendants of the hybrid offspring eventually flew to mainland Italy, where the locals caught them and thereafter devoted themselves to the rearing and selling of canaries. Another account involved a Dutch sailor who managed to catch a few pairs 'which once introduced to his country, multiplied very easily in captivity'.

As Hervieux knew, the Elba shipwreck story was a fantasy. Like most fantasies, however, it does contain a grain of truth, which is that there *are* some very odd canary-like birds on Elba. The mystery of their identity wasn't resolved until almost 400 years after Todi's account, when two Italian ornithologists rediscovered a painting of the 'Elba canary' in the Royal Library at Windsor Castle. Painted around 1630 by the talented Vincenzo Leonardi, the picture shows not a canary but a completely different species, the citril finch. Leonardi's painting is so accurate that we can also see this isn't a typical citril finch (like those which live in the mountains of mainland Europe), but the rather distinctive Corsican race of citril finch. In other words, the hybrid canary of Elba is a red herring.[9]

Olina did more than borrow the Elba canary story from Valli da Todi. Their two books are suspiciously similar in every other respect. The similarity led later scholars to think that Olina was a crook: the plagiarist's plagiarist, they called him. As one writer pointed out, 'Plagiarism was common . . . but Olina easily stands first as an adept at this practice.' On the face of it, Olina's rehash of Valli's book looks like a blatant case of unacknowledged borrowing. It wasn't Olina who was the crook, however, but one of his powerful patrons. His magnificent book was commissioned by Cassiano dal Pozzo employed at the court of Pope Urbano VIII. Cassiano arrived in Rome in 1618 keen to carve a career for himself. He used his immense diplomatic skills to make Rome the centre of European culture and arts, accumulating vast collections of historical, artistic and natural history works as well as commissioning new ones. He soon became a key player in the new wave of scientific objectivity of which Galileo was a part and which demanded the accurate portrayal of nature. An astute observer and bird-fancier, Cassiano used the same care in describing plumage as he did when recounting what the ladies at the European courts were wearing.[10] It was he who arranged for Giovanni Pietro Olina to prepare the definitive book on bird keeping for the Accademia dei Lincei (Academy of Lynxes), Italy's premier cultural organisation founded in 1603, whose aim was to use its sharp eyesight to penetrate the secrets of nature.

Twenty years earlier, in 1601, Antonio Valli had published his book *Il Canto degli Uccelli* (The Song of Birds) – but few copies were printed and it might have disappeared without a trace had not Cardinal Del Monte, one of Cassiano's protectors, owned one and liked it. Cassiano and Olina realised that Valli's book could form the basis for a more comprehensive and elaborate volume. Accordingly, Valli's original text was tarted up by adding a few classical references and some scientific facts, and one of the original illustrators, Antonio Tempesta – now somewhat aged – was wheeled out of retirement to retouch and improve his original plates. There were new illustrations, too, notably by the great natural history artist Vincenzo Leonardi, who painted the 'Elba canary' found in the Windsor Castle library centuries later. After months of preparation, on 15 August 1622 Cassiano presented *L'Uccelliera* to the Accademia with the following message: 'I send you this bird book produced by one of my affiliates as a tribute of my respect, as evidence that the information I gather with limited effort and money can contribute towards this field [of science].'

Cassiano hadn't gone to all this trouble out of mere scientific curiosity or academic altruism. The truth was that he was desperate to get himself elected to the Accademia. He had written much of the new book himself, but resisted the temptation of adding his name to Olina's as co-author because he was terrified that the Inquisition and the Church might misjudge his motives. It was distinctly dangerous to be seen to be overtly ambitious or indeed overly curious about the natural world in the Italian royal court, as Galileo was to discover to his cost a decade later. Olina was the fall guy and if things turned sour it would be he and not Cassiano who got nailed. But in fact they both emerged smelling of roses. *L'Uccelliera* was a stunning success for Olina, running through numerous editions and the Accademia eventually acknowledged Cassiano's monumental contribution to science and art by allowing him to join their ranks.

Origins

The wild canary comes, as its name implies, from the Canaries – best known now as the holiday islands of Tenerife, Lanzarote and Gran Canaria. Canaries also occur on the islands of Madeira and the Azores, where they were probably deliberately introduced. Originally referred to as 'Canary-birds', they were named for the islands rather than the other way around. The islands were previously known as 'Canaria' in recognition of the large dogs (*Canis*) that the occupants owned in ancient times. In the first century AD, the Romans referred to the archipelago as 'The Fortunate Isles' on account of their benign climate – lying as they do at the same latitude as Cairo, but with the benefit of ocean breezes and good rainfall. After their brief encounter with the Romans the islands and their inhabitants enjoyed more than 1200 years of blissful obscurity before they were rediscovered in the mid-1300s. Word got around, and at the turn of the century the French adventurer Jean de Bethencourt arrived and began the bloody business of conquest, subjugation and settlement in the name of the King of Spain. Legend has it that Bethencourt was so impressed by the song of the caged canaries owned by the local people that he took some back with him as gifts to the court of Castille.[11] He later gave one to the Queen of France, Isabeau of Bavaria, wife of Charles VI – thereby initiating the French royal family's protracted love affair with the canary. In the hundred years following Bethencourt's arrival on the Canaries, having succeeded in completely eliminating the islands' human population, successive Spanish expeditions began to plunder their birds.

The Spanish were astute traders and maintained a shrewd monopoly on canary birds, allowing only live males to be exported to Europe. Louis XI, King of France from 1461 to 1483, famous for founding three universities, is said to have been particularly fond of canaries and other songbirds. He apparently bought *chardonnerets, lignots, verdiers* and *pincons* (goldfinches, linnets, greenfinches and chaffinches) by the hundreds, and canaries – by the dozen. As transport increased, canary birds were imported more frequently into mainland Europe and by the middle of the 1500s the export of canaries was a well-established and highly profitable business. A single shipment in 1546 comprised twenty-five dozen birds from Gran Canaria. Overall, the numbers exported each year must have amounted to thousands and this vigorous trade in wild birds continued unabated for at least a further 200 years. *The London Gazette* in 1685 reported the arrival of 700 canary birds from Canary. In view of what happened elsewhere under similar circumstances (see Chapter 9) and the relatively low numbers of wild canaries available – a world population of no more than 160,000 pairs in total – it is remarkable that they were not exterminated in the frenzy to supply the demanding European markets.[12]

One of the great polymaths of the Late Middle Ages, William Turner (c. 1510–68), famous for his *Herball* and revered as the father of British botany, but also as an ornithologist, provides one of the earliest records of canaries in Europe.[13] Discussing the canary grass *Phalaris canariensis*, on whose seeds the birds were maintained, he said, '. . . for they that brought Canari burdes out of Spayn bought of the sede Phalaris also to fede them with'. Even though he had never seen one, Turner's close friend, the Swiss naturalist Conrad Gessner, included the canary in his animal encyclopaedia, *Historia Animalium*, published in the 1550s, on the strength of Turner's description. Gessner wrote, 'It is sold everywhere very dear, both for the sweetness of its singing, and also because it is brought from far remote places, so that it is wont only to be kept by nobles and great men.'

Spain at this time had an empire on which it was said that the sun never set. But late in the sixteenth century the sun did start to dip below the horizon and one of the first indications of Spanish decline was its loss of the canary monopoly. Until this time anyone wanting one purchased an imported, wild-caught male bird from a Spanish dealer. But the occasional female must have slipped through the nets, despite the traders' best efforts. Having got their hands on females, it was the Germans, in their typically organised way, who set about breeding them and who, by avoiding all the middlemen and the transportation costs, eventually undercut the Spaniards. As soon as they could reliably rear canaries, the German bird enthusiasts either wittingly or unwittingly started the process of artificial selection – partly on colour, but primarily on the canary's song. Breeding canaries for song was a monumental commercial success both within Germany and abroad, generating a trade that persisted for 350 years. Canaries were marketed across the length and breadth of Europe and Hervieux recounts how when German dealers brought birds to Paris twice a year in spring and autumn, scuffles broke out as people rushed to buy them.[14] He had a jaundiced view of the German traders:

> When you go to ask Questions of them concerning their *Canary-Birds*, or other such like Affairs, without Buying any thing, they give you a very bad Reception, and in short, they use you very roughly . . . but as soon as you show them that precious Metal, without which the most Ingenuous Man is not valu'd, I say, when they perceive you come to Buy some of their *Canary-Birds*, they receive you very courteously, and in their broken Language express themselves very much your Humble Servant . . .

In England the naturalist John Ray, writing in 1678, reported that the canary birds imported from Germany 'in handsomeness and song excel those brought out of the Canaries'. The birds were so popular that German entrepreneurs set up what we might now describe as canary-leasing businesses in which, for an appropriate fee, someone would deliver a singing bird, turn up every day to feed and water it and, when it stopped singing, replace it with another. You could even specify whether you wanted a daytime or a night-time singer.[15] The other thing the Germans did, albeit unwittingly, was to 'domesticate' the canary.

Their Parentage

In each case starting with a single species, humans have created more than 300 breeds of domestic pigeon, over a hundred dogs, dozens of breeds of cats, mice, sheep, pigs and cattle, and some seventy breeds of canary. Darwin was overwhelmed by the success of farmers, stock-breeders and pet owners in producing these different breeds. He exploited their success to explain the mystery of mysteries: the origin of species. Using man's artificial selection of varieties as an analogy for natural selection was a stroke of genius. This was the heyday of animal breeding, and everyone in Victorian England and elsewhere in Europe was well aware of the 'improvement' that had been made in the different breeds of farm animals and pets by artificial selection. Domestication is the largest experiment ever undertaken, but prior to Darwin no one saw any value whatsoever in trying to understand its results. It took an immense effort, but in essence all Darwin had to do to make his point about natural selection was to draw a careful parallel between the ways the hand of man and the hand of nature selected the fit and weeded out the rest.[16]

What Darwin did not bother to pursue was the question of what predisposed particular species to become domesticated in the first place. This was left to his younger scientific cousin, Francis Galton – dilettante and brilliant maverick – who recognised that to have any chance of becoming domesticated, animals must fulfil six criteria.[17] They had to be hardy, comfort-loving, useful ('to the savages'), free-breeding, easy to tend and they had to have an inborn liking for people.

By 'comfort-loving' Galton simply meant that they should settle down and adjust to captivity, and not constantly try to escape. Animals had to be useful, either as a source of food (like chickens, sheep, pigs and cattle), as protectors (like dogs), or as hunting assistants (dogs again). Unless animals bred freely in captivity then, like many zoo animals, they would never become domesticated. Galton's idea that animals should be easy to tend has been paraphrased by others who noted that to become domesticated animals must be able to thrive on neglect. More generally, what Galton had in mind was that livestock should be versatile in their diet and show a tendency to remain together, allowing themselves to be herded and controlled. The final attribute, a liking for people, refers to the idea that animals should be socially subservient and accept humans as their natural leaders.

Galton summed up his treatise on domestication by saying, 'It would appear that every wild animal had had its chance of being domesticated, that those few which fulfilled the above conditions were domesticated long ago, but that the large remainder, who fail sometimes in only one small particular, are destined to perpetual wildness so long as their race continues. As civilisation extends they are doomed to be gradually destroyed off the face of the earth as useless consumers of cultivated produce.' In addition to his essay on domestication, Galton is remembered for two things. He discovered that human fingerprints were unique and could be used to identify individuals, and he was responsible for coining the term and starting the study of eugenics – the idea that humans, like domestic animals, could be selectively bred and 'improved'. The way Galton introduced this idea seemed more than reasonable: 'Man is gifted with pity and other kindly feelings; he also has the power of preventing many kinds of suffering. I conceive it to fall well within his province to replace Natural Selection by other processes that are more merciful . . .' This syrupy language masked a programme to promote naturally gifted members of society, the genetic elite – like himself – at the expense of the rest.

Leaving aside for a moment the fact that, through no fault of their own, canaries became mixed up in the eugenics movement, how applicable are Galton's six attributes to the wild canary? To have survived the boat journey from their native islands to the shores of mainland Europe and beyond, the original wild canaries must indeed have been very hardy. One of the most important consequences of domestication was the accidental – and brutal – selection of healthy, vigorous birds. Judging from the numbers of pages devoted to remedies for various 'direful maladies' in early books, captive canaries must have been horribly susceptible to disease – in part because they were often kept in squalid conditions. This shouldn't come as a surprise, given that their eighteenth-century owners knew precious little about maintaining themselves in a sanitary condition. Paris at this time was described as stinking with filth and during the 1700s the entire country was ravaged by wave after wave of plague.[18] Hervieux, however, knew how important cleanliness was in keeping his birds in good health, as indeed did another nation of canary fanatics, the Dutch, whose national obsession with cleanliness is thought to have contributed to their particular success with breeding canaries.[19]

The concept of 'comfort-loving' canaries is a mystery. Everyone who has taken wild canaries into captivity in the last hundred years or so and written about their experience has commented on how reluctant their birds were to settle down – they never become tame. Is this surprising? Should we expect any wild animal to recognise the benefits of captivity and sit back and enjoy its numerous advantages? Remarkably, that is just what some species of birds do and it was precisely this characteristic that made the goldfinch the most popular native cage bird in Europe. In his handbook on birds published in 1676, Francis Willughby described the goldfinch, thus:[20] 'They are of a mild and gentle nature, as may even hence appear, that presently after they are caught, without using any art or care, they will fall to their meat and drink, nor are they scared and affrighted at the presence of man, as to strike their bills and wings against the sides of the cage, as most birds are wont to do.' The wild canary was the antithesis of this.

Canaries were never 'useful to the savages' in the way that pigs, sheep or dogs were: their only value was their song. They were a luxury and on the Canary Islands they were kept because people enjoyed listening to them. But once the first representatives of European royal courts appeared there, canaries suddenly became extremely useful as status-enhancing souvenirs.

Wild canaries in captivity certainly were *not* free-breeding, and their reluctance to reproduce is closely linked with their resistance to settle down. Those who keep wild canaries have complained about the extraordinary difficulty of getting them to breed and have been amazed that anyone in the past ever succeeded.[21]

Canaries, like many other finches, are very easy to tend and can survive on little more than dried grass seeds, a bit of grit (to grind up the seed in their gizzard) and water. The reason other fine singers like nightingales, blackbirds and larks never became domesticated is that they require much more sophisticated diets than the canary.

It is far from clear whether wild canaries ever had 'an inborn liking for humans', although the tameness of domesticated birds was certainly one of the attributes that later made them popular.

In summary, wild canaries fail miserably on two of Galton's six criteria: they are neither 'comfort-loving' nor 'free-breeding'. How, then, did canaries become domesticated? The most likely explanation is that although the majority of exported canaries were taken from the wild as adult birds, some at least must have been young birds, taken from the nest and reared by hand. This was common practice on mainland Europe – Hervieux gives detailed instructions on how to do it with all sorts of bird species[22] – and it meant that young canaries, like Nicolai's *Gimpels*, grew up believing their human owner to be their parent. The local people on the Canary Islands must have known, even by Jean de Bethencourt's time, that hand-reared birds sing more and survive better because they don't thrash around their cage trying to escape. In addition, tame canaries would also have been much more valuable as trade objects than neurotic ones. All the recent writers describing bad experiences with wild canaries in captivity have had wild-caught adult birds. Hand-reared canaries would have been comfort-loving and much more likely to breed in German aviaries.

Being bigger, sexier, dafter and less colourful than your wild counterparts is an inevitable consequence of domestication.[23] Domesticated animals are invariably larger than their wild ancestors, partly because breeders have artificially selected them for size, but also because captive animals overeat and under-exercise. But why sexier? Because by definition domestication selects for animals that will breed in captivity. Attempts to study the process of domestication with wild-caught rats, for example, have shown that only a minority of individuals ever reproduce in captivity. Those that do are the more sexually motivated, with the result that domesticated animals tend to breed more frequently than their wild ancestors. The jungle fowl, progenitor of the domestic chicken, becomes sexually active at one year old, but chickens are ready to breed at just three months old. In a year a typical jungle fowl would lay one or two clutches of a dozen eggs while their domesticated descendants pump out over 300 in the same period.

Dafter? Think of sheep – sadly lacking in common sense, some of which has been lost because breeders positively wanted more docile and manageable stock, but also because the intense selection pressure provided by predators like wolves in the wild has been removed. Domestication is little more than survival of the dumbest – under the guiding hand of man. But stupidity is not simply a product of the genes. It is a direct consequence of an institutionalised life.[24] Domestic life may be bliss but it is also unstimulating, and an unstimulated brain – even a bird brain – is only half a brain. New research has shown that even wild-caught birds are less intelligent than their wild counterparts. The bird's brain is a powerhouse of neuronal rejuvenation. Nerve cells come and go with unbelievable frequency, and in response to the cognitive challenges of finding food, dodging sharp-clawed predators, learning to sing and selecting a partner. Without these everyday challenges, the rate of neuronal renewal is depleted and the brains of captive birds actually decrease in size. These recent findings have shaken the very foundations of neuroscience because it was previously assumed that, as in ourselves, the brain an animal is born with is the one it is stuck with. But birds seem to be unique in their dynamic pattern of cell replacement, allowing them constantly to update their faculties with no overall increase in brain tissue or metabolic costs.[25]

Metamorphosis

Of all the traits shown by domesticated animals, the loss of colour is perhaps the most obvious. Indeed, we often recognise an animal as being domesticated simply from its colour: white rats, white mice, white cattle, white rabbits and white cats are all missing their normal dark pigments. Sometimes the loss of pigment is total, resulting in pink-eyed individuals – albinos. If the loss is limited to the skin (including the feathers in birds) the result is referred to as 'leucism', a white individual with dark eyes. Sometimes only one of several pigments is missing: wild budgerigars are green and, just as with paints, this results from a mixture of blue and yellow colours in the feathers. An absence of yellow results in the familiar blue budgerigar; if the blue is absent, we get a yellow bird. Often the colour difference between the wild and the domesticated forms is due to a single gene and, remarkably, in a few cases biologists have even located the gene among the tens of thousands, and figured out exactly how it creates its colour-diluting effects. In the canary's case, Duncker's research showed that the canary's transformation from green to yellow occurred as a result of the change in at least three genes and the concomitant loss of a dark pigment, melanin, from the feathers.

Duncker's genetic knowledge and ability to conduct and inter-pret his canary experiments were largely self-taught. Genetics was still a very young discipline when he finished his university studies in 1905. There was a huge amount of interest in patterns of inherit-ance at this time, especially among researchers in America and Britain. We know that Duncker was an exemplary teacher and that he kept up to date with new developments in biology, including genetics. His understanding of genetics – particulate inheritance subject to the laws of probability – must have been greatly facilitated by his aptitude for physics and mathematics, which he also taught. Duncker had probably inherited his Darwinian views as a student from the pervasive Ernst Haeckel without, it seems, accepting

Haeckel's Lamarckian belief in 'progress'. Instead, Duncker focused on pure Mendelian inheritance; inheritance devoid of any outside influence or purpose. Such was Duncker's enthusiasm for Mendelism that he taught it at school, demonstrating its truth by conducting simple genetic experiments for his pupils. From the books and papers Duncker cited in his own publications it is clear that he watched the field of genetics grow from its uncertain and controversy-ridden beginnings to the more mature and forward-looking period around 1920. By the time of his meeting with Reich, Duncker knew and understood all that there was to know about the way traits, such as plumage colour, were inherited.

The manner in which the canary's green-to-yellow transformation came about resembles the situation in which a family has had a much-loved but very familiar painting in its possession for generations and barely given it a second glance. Then one day a patch of paint flakes off, revealing that underlying the well-known image there may be another much more attractive one.

We know that there were yellow canaries in William Hogarth's day in the early 1700s. At this time England was a country of pick-pockets, press-gangs, wig-wearing workers, urban filth, rural poor, child chimney sweeps, slaves, grave robbers, public hangings, gin mania, cockfighting, Fanny Hill and rickets, as well as enthusiastic bird keepers. In London alone there were an estimated 200,000 canaries and people craved information about their pets and how to care for them.[26] Accordingly, several books on bird keeping were produced in the early 1700s – many published anonymously and most pirating each other's information. In 1718 even Hervieux's book was translated into English – possible only because of the Treaty of Utrecht, which ended the War of the Spanish Succession in 1713 and also marked the end of decades of animosity between the French and the English, and opened up trade and cultural exchange between the two countries. Wonderful as Hervieux's book was, it had no pictures and was eclipsed in England by another, using a brilliant marketing

ploy for the first time – a bird book with *colour* plates. Eleazar Albin's
three volume *A Natural History of Birds* was published in 1731, 1734
and 1738 and its lifelike colour images must have made it irresistible
to better-off bird enthusiasts.[27] Born in Germany, Eleazar Weiss
moved to England in 1707 in his mid-twenties, changed his name to
Albin and set up shop near the Dog and Duck public house on
Tottenham Court Road, London, as a professional watercolourist
and art teacher. He and his daughter Elizabeth drew and coloured the
plates for their book, priding themselves on having drawn the birds
from life (which sometimes meant from death) – including a very
special canary.

Albin describes the various types of canary then available, including all-yellow birds, but the one he illustrated was white. He knew
what he was doing, for although white canaries were not uncommon
in his native Germany, they were extremely rare in England and he
knew that fanciers loved unusual birds. His three-volume work
depicted several extraordinary birds, including a black bullfinch and
a strange hybrid described in the next chapter, and he went out of his
way to find such oddities. In the preface to the first volume he wrote,
'I will be very thankful to any Gentleman that will be pleased to send
any curious birds (which shall be drawn and [en]graved for the second
volume, and their names shall be mentioned as Encouragers of the
work) to Eleazar Albin.'

All but one of the plates in Albin's book were drawn and coloured
by himself or his daughter. The exception was the 'Red Linnet Cock',
which was coloured by Albin's eighteen-year-old son, Fortin. The
reason why Fortin was allowed to colour this particular plate may
have been because he helped to make the red pigment. In preparing
his colours, Eleazar Albin sometimes used what today at least would
seem like novel ingredients and it was said that 'for his reds he washed
dried vermilion pigment in four waters and then proceeded to grind
it in boys' urine three times, then gum it and grind it in Brandy
wine'. Presumably Fortin supplied the urine, but it is not clear why

Elizabeth's wouldn't have done just as well – unless a touch of testosterone added a certain *je ne sais quoi.*

Albin's book was a landmark in bird publishing, although not everyone rated it highly. Duncker's contemporary Erwin Stresemann,[28] director of the Zoology Museum in Berlin, considered Eleazar's and his daughter's talents very limited and described Albin as being 'as clumsy with the pen as with the brush'. Stresemann was particularly critical because Albin, knowing nothing about birds, had lifted his entire text from earlier writers. To be fair, so had just about everyone else and at least Albin had the grace to say from whom the information was stolen.

Because plagiarism was so pervasive, paintings provide a much more reliable record of the canary's change from green to yellow than written accounts. But the search for pictorial records was to be undertaken not in England or France but in Germany, the crucible of canary creation. Erwin Stresemann was the first to recognise the scientific value of bird illustrations after discovering in 1922 that his museum's vaults contained some beautiful canary paintings dating back to 1600.

Stresemann's interest in art and canaries typified his all-embracing approach to ornithology and in his mid-thirties he single-handedly, but with great determination, began the modern revolution in bird knowledge.[29] Before 1925 the study of birds had been the domain of amateurs and dilettantes. Through his astute appreciation of what was worth pursuing, Stresemann pushed ornithology to the vanguard of mainstream zoology. His own ornithological work at the Berlin Museum spanned almost every aspect of zoology, from migration and anatomy to behaviour and evolution, including the process of domestication. Stresemann liked Duncker's work, for Duncker too was forging novel links and squeezing knowledge sequestered by generations of bird keepers into a scientific mould. Stresemann was also sympathetic to Duncker's interest in cage birds, probably because Stresemann's own career had begun when, at the age of

sixteen, he published his first paper describing how he had successfully created a hybrid between two cage birds, goldfinch and redpoll.
Unlike most of his colleagues, who ignored domesticated birds
because they were not 'real', Stresemann appreciated that their story
could reveal something about the origin of species, which, despite
the title of Darwin's book published sixty years earlier, was by no
means resolved or understood.[30]

The paintings Stresemann discovered in the Zoology Museum's
vaults were by Lazarus Röting, whose entire collection of natural
history art had been bound together for safe-keeping and entitled
'Theatrum Naturae' by his nephew in 1615, the year after Röting
died. So named because his desperate parents hoped for a miracle,
Lazarus was born in 1549 with glassy-bone disease, *osteogenesis
imperfecta*, the same condition that later afflicted the painter
Toulouse-Lautrec. Like Lautrec, Lazarus Röting's legs were damaged
before he even set foot into the world. He was the runt in a brood of
fifteen children and remained a cripple throughout his life. His father,
who taught languages and theology at the Gymnasium in Nuremberg,
tutored his housebound son at home and Lazarus in turn taught
himself to paint, concentrating his efforts on local wildlife and cage
birds, imbuing his animals with a sprightliness he himself lacked.
Being housebound may have been no bad thing for Lazarus, since
plague was rampant and the local population was given to violent
outbursts of both anti-Semitism and witch-hunting. Röting's beautiful paintings reflect none of these troubles and are so accurate that we
can safely assume they provide an honest record of what he saw. He
painted several canaries, all of them typical wild, green birds – except
for one that has white wings and a few yellow body feathers.
Stresemann was convinced that this particular painting, completed
around 1610, caught the very beginning of the canary's transition
from green to yellow.[31]

If Röting's painting signalled the beginning, when was the process
complete and the first all-yellow canary produced? Searching among

the book collections of Berlin's libraries, Stresemann made his first find: an article written in 1702 by Rosinus Lentilius, a physician from Nördlingen near Nuremberg in southern Germany, which describes white canaries as commonplace. Lentilius goes on to say that one particularly clever weaver boasted that he could breed canaries of any colour – but refused to reveal his secret.

Any colour? Tantalising, but too vague. Stresemann then came across a report of all-yellow canaries in the 1570s from Francisco Hernandez, chief physician to Philip II of Spain who had sent him to Mexico to document its natural resources. Hernandez said that the canaries in Mexico were *tota lutea* – all yellow – but Stresemann recognised immediately that this was a case of mistaken identity. These yellow birds were another species altogether. Eventually, closer to home Stresemann discovered the account by Lucas Schroeckius of Augsburg, another physician, who in 1677 described the first completely yellow canary. At this time such birds must have been scarce and so incredibly valuable that they were unknown in England and the likes of Samuel Pepys had to make do with green ones.[32]

Stresemann concluded that it had taken German fanciers sixty or seventy years – from Röting's first flecks of yellow in 1610 to Schroeckius's all-yellow birds of 1677 – to change the canary from green to yellow.

But this wasn't the whole story. Paintings of part-yellow canaries were in existence some thirty years before those of Röting. They were included in a thirty-three volume encyclopaedia-cum-diary, *Thesaurus Picturarum*, produced by the Protestant cleric Marcus zum Lamm who was employed at the royal palace in Heidelberg. Lamm, a direct contemporary of Röting, devoted no less than three volumes to birds, each of which was illustrated in colour by unknown artists. There are four paintings of canaries, two wild green birds (which, Lamm noted, bred in captivity) and two birds with bright-yellow breasts, all thought to have been completed about 1580. Lamm says that these unfamiliar part-yellow canaries had come from Tyrol and that he was unsure

how to categorise them. The very fact that Lamm had access to such
birds confirms that, just as elsewhere, canaries were owned only by
the aristocracy. The paintings he commissioned suggest that as early
as the 1580s selective breeding among German enthusiasts was starting to produce interesting results. His extraordinary encyclopaedia
has been one of Heidelberg University's most treasured possessions
ever since Lamm's death in 1606. Stresemann got to hear about it
through his student Ernst Mayr who was searching for historical
records of the serin (a close cousin of the canary, which was expand-
ing its range across Europe) as part of his doctoral studies. A young
historian, Albrecht Schwan, drew their attention to Lamm's encyclopaedia several years after Stresemann had written about Röting's
yellow-flecked canary. Schwan went on to study Lamm's works for
his own Ph.D., which he completed in 1926, but it was not until 2001
that the illustrations and other material in these volumes became
generally available.[33]

Compared with the exquisite paintings by Röting, those in Lamm's
encyclopaedia are rather crudely executed, but the colours are extraordinarily brilliant. This is especially true of the yellow plumage on the
two canaries. Intense yellow was a particularly difficult colour to
achieve in Lamm's time and the most readily available source of
pigment came from the berries of plants such as buckthorn. It is no
coincidence that the yellow pigments in the berries – substances called
carotenoids – are the very same ones which in real life give canaries
their yellow colour, the only difference being that carotenoids were
even more effective in feathers than they were on paper.[34]

That partly yellow canaries existed by 1580 is clear from Lamm's
encyclopaedia, but it also turns out that there were entirely yellow
canaries sooner than Erwin Stresemann or anyone else thought. The
talented German bird artist Johann Walter painted entirely yellow
and white canaries in 1657,[35] twenty years earlier than those recorded
by Lucas Schroeckius. Taken together, the pictorial records of Lamm
and Walter indicate that it had taken around a century – one hundred

canary generations and three generations of canary breeders – to complete the green-to-yellow transition. Had he lived long enough to learn about the discovery of Lamm's and Walter's canary paintings, Duncker would have been delighted; a century was exactly the kind of time-frame he imagined for the canary's transition. Even Stresemann's estimate of sixty to seventy years was reasonable. Duncker knew from his breeding experiments with Reich that the canary's change in colour could not have occurred more rapidly.

In fact, the yellow canary *could*, in principle, have been produced almost at a stroke, exactly in the manner Davenport imagined, because this is precisely what happened with another bird. The greenfinch, long a favourite cage bird in Britain, superficially resembles the wild canary in its colouring: a green bird with some grey, brown and yellow feathering. But a perfectly clear yellow greenfinch also exists, which bird keepers refer to as a 'lutino', a single gene mutation that, like the albino human, lacks any melanin whatsoever and has pink eyes. The only difference is that in the absence of melanin, the greenfinch's ground colour is yellow rather than white. Very occasionally lutino greenfinches occur in the wild; they are very rare – perhaps less than one in a million – but as there are some 10–20 million pairs of greenfinches in Europe at any one time, seeing one is still somewhat more likely than winning the lottery. A lutino greenfinch is startlingly conspicuous and cage bird enthusiasts on seeing one would have made every effort to catch it. Owning one would have conferred status and breeding from it even more so. And that is exactly what happened. Lutino greenfinches were captured, paired with normal greenfinches and their offspring – which look exactly like normal greenfinches but carry the recessive lutino genes – were mated back to the lutino bird. In this way and with a great deal of patience, starting in the 1940s breeders captured the lutino genes and developed a strain of pure yellow greenfinches in a mere twenty years.[36] The speed with which this was achieved was largely due to the existence of a completely yellow single gene mutation; but also, by the mid-1900s the understanding of inheritance was much greater than it was in the 1500s and 1600s when breeders were trying to create a yellow canary.

If a perfectly yellow mutation can arise spontaneously in the greenfinch, why couldn't the same thing happen among wild canaries? The answer is that it could, but it didn't, probably because compared with the greenfinch the world population of wild canaries is so small. With his incredible biological insight Darwin was well aware of this and wrote in the *Origin of Species* that the smaller the population, the less likely it was that 'a well marked variety', (like a lutino) will crop up. As far as anyone is aware, a wild yellow canary mutant has *never* been recorded. As a consequence bird enthusiasts had to produce yellow canaries the hard way.

6

Domestic Life and Death

Whoever has some experience with breeding canaries will know that, if a white Nightingale (a male) is mated to an ordinary Nightingale (a female), there will result no other than Nightingale of ordinary colour in the first year. Next year, however, if one mates a young female of this offspring again with the white male, the white colour will appear in some of their offspring, and in the third year (after mating white with white) no others than white birds will result.

BARON VON PERNAU, *Angenehmer Zeit-Vertreib* (1716)

Over beer and more cigars, Duncker asked Reich how the various different canary breeds had arisen. Like most fanciers, Reich believed that much of the variation in canary size, shape and song occurred as a result of crossing the original wild bird with other finches, but Duncker wasn't convinced. After all, Darwin had shown that the entire range of pigeon varieties, infinitely more diverse than the canary, could be traced back to a single ancestor, the wild rock dove *Columba livia*. Duncker could see no reason why the same shouldn't be true for the canary.

But mapping out the canary's family tree and pinpointing the emergence of each breed proved to be more difficult than he anticipated. Canary strains were often much less distinct than pigeon breeds and because they arose only gradually there was never a

precise moment when a canary fancier could announce the creation
of a new variety.

The process of domestication was Darwin's model of evolution in
nature. The development of different pigeon or canary varieties was
equivalent to a single species giving rise, over time, to a number of
new ones radiating out to occupy a variety of niches. This is exactly
what Darwin imagined had happened on the Galapagos Islands with
the ground finches in a process ecologists now call 'adaptive radiation'. The Galapagos were also where species suddenly seemed less
like fixed entities created by God and more like something mutable.

Darwin visited the Galapagos on his *Beagle* voyage in 1835 and, just
as he had done elsewhere, he made collections along with other crew
members, including Captain FitzRoy, of the local wildlife. Darwin
had little idea that the various small birds they so carefully killed,
stuffed and labelled would help to turn the world upside down. Only
after he had returned home and was ruminating over his Galapagos
experiences did the full significance of these little birds become clear.
They differed so much in appearance, especially in the size of their
beaks, that Darwin struggled to identify them and he labelled them
variously as finches, wrens and gross-beaks. Once he was back in
London, he passed his study skins on to John Gould, an up-and-
coming ornithologist and artist who, unlike Darwin, recognised
immediately that, despite the differences in their beaks, the birds
were very closely related. More important, Gould realised that they
were unique and comprised an entirely new group of twelve species
of ground finch. None the less, when Darwin came to write his
*Journal of Researches into the Natural History and Geology of the
Countries visited during the Voyage of H. M. S. 'Beagle'* (1839) he was
still dithering about whether the finches really were separate species
or merely varieties of a single species. As he deliberated, an incident
came to mind in which the resident Spaniards on the Galapagos told
him that they had only to look at the shell of a giant tortoise to be
able to say which of the various islands it came from. The tortoises

and finches had more in common than Darwin thought. Each group of species had originated from a single mainland ancestor and after arriving on the Galapagos aeons earlier eventually spread to the separate islands where the different environmental conditions caused them to diversify.[1]

Now imagine a rerun of the Galapagos finch saga in sixteenth-century Europe. In this scenario the finch ancestor resides in Germany, having been brought there from its native Canary Islands a few centuries earlier. The German breeders 'improved' the wild canary in captivity by artificially selecting birds on the basis of their song. Unwittingly, though, they also did two other things. First and rather obviously, they selected canaries for their ability to reproduce in captivity. Second, by repeatedly breeding those few individuals that accepted captivity, they released some of the heritable variation that had previously lain unexpressed in the wild canary's genes. In focusing their artificial selection efforts entirely on the birds' song, the Germans, initially at least, ignored the occasional white feather or clear yellow cap. Over time, more and more variations in colour, size and shape appeared through the chance combination of rare genes and some German fanciers decided to perpetuate this variation by deliberately choosing to breed from birds of particular colours – like those illustrated in Lamm's book. Nonetheless, the market was primarily for singing birds and the quality of the birds' song remained the main priority.

By the early 1700s when Hervieux wrote his famous monograph, canaries showed quite a lot of variation in colour but there were few, if any, truly distinct varieties. Once the English got their hands on German birds in the early eighteenth century, however, all that changed. It was equivalent to the first finch ancestor arriving on the Galapagos. The analogy goes even further: the different towns in England – Norwich, Manchester, London – were like the different islands of the Galapagos archipelago. Each one imposed a slightly different selection regime on its canaries, pushing them out along different branches of an evolutionary tree.[2]

In plotting out the canary's evolutionary history Duncker struggled to decide which of the early authorities he could trust. So many
of those who wrote about canaries cribbed their information from
earlier accounts that it was extraordinarily tricky to pinpoint the
origin of information, let alone a particular breed. The key period for
the canary's domestic evolution, Duncker soon realised, was the
century between about 1720 and 1820. Had Eleazar Albin been a
better observer, his *Natural History of Birds* might have helped, but he
merely noted the presence of six types, similar to the six listed by
Thomas Hope, whose little book, *The Bird Fancier's Necessary
Companion* published in 1762,[3] included a description of what he calls
the '*Spangled Sort*'. This is probably the same as what Albin calls 'the
most beautiful feathered bird' and what we now recognise as the
Lizard, and probably the very first established variety.[4] Hope captured
the essence of the canary breeders' mentality, and their obsession with
scarcity, when he wrote, 'Therefore, For the meer uncommonness
only of the Thing it is, that the Black Tails, and Cap'd Birds, are most
esteemed . . . So that this is nothing but meer fancy, because Birds
with White Tails, & no Caps, are so Common.'

Meer fancy. The birds whose particular traits breeders like Hope
actively chose to perpetuate were *fancy* birds; these were the selected
lines, the future gene stock of particular breeds. The rest were
'common or gay' birds – mongrel canaries, whose carefree breeding
included no future vision, no fantasy. A bird-fancier was someone
who bred fancy birds; the terms 'fancy' and 'fantasy' have the same
root, reflecting the fact that fanciers focused their efforts on creating
something they had in their minds' eye.

Thomas Hope also reported how 'Some Canary Birds are longer
from Head to tail, are taller, & have blacker Legs than others . . .
Of these, The Best Sort, wither for Singing, or Breeding, are the
Tallest, and of Near a Span Long, from the Bill, to the end of the
Tail . . .' A span was the distance from the end of the thumb to the
end of the little finger, a length of about nine inches. These long,

thin canaries, which eventually gave rise to a number of breeds, including the Yorkshire Fancy, had their origins in Flanders and were apparently brought to England by French Protestants known as the Huguenots.

The Huguenots had a marked effect on English culture including, legend has it, its bird-keeping culture. Between the sixteenth and eighteenth centuries huge numbers of Huguenot refugees fled from the Continent to England to escape persecution by their Catholic countrymen. The oppression began in France when the first Huguenot was burnt at the stake in 1523, and once it started the ethnic cleansing was relentless. When Louis XIV legalised massacre through the Edict of Nantes in 1685, 250,000 Protestants promptly fled to England, Holland and North America. Recognising the contribution that skilled Flemish and French craftsmen could make to the country's economy, England's Protestant rulers welcomed them. Of course, the asylum seekers were opposed by local artisans who felt threatened by this influx of foreign talent, but the refugees richly repaid England for its hospitality, improving virtually every manufacturing trade, including the making of paper, glass, cloth as well as carpentry, hat-making and horticulture.

The most successful of the Huguenot immigrants were cloth makers, especially silk and wool weavers, who settled in the city of Norwich, and further north in Yorkshire and Lancashire, bringing their birds with them. Weaving was a cottage industry and working at home was entirely compatible with canary keeping. The birds' song kept the weavers entertained as they worked, just as the radio does in today's factories. The birds' breeding output provided some supplementary, untaxed income; and the quality of their birds became a point of competition for men when they were not at work.

Some of the opposition the English felt towards these immigrant workers – strangers, sharpers and intruders, Hoghen-Moghens, 'Hugunots' and shoemakers – was expressed in an anonymous political poem published around 1708, which starts:

CANARY BIRDS NATURALISED IN UTOPIA
A Canto

In our unhappy days of Yore,
When foreign Birds, from German Shore,
Came flocking to Utopia's Coast,
And o'er the Country rul'd the Roast:-
Of our good people did two-thirds
So much admire Canary birds
For outward Show, or finer Feathers
Far more regarded than all others.
We bought 'em dear and fed 'em well,
Till they began for to rebel.

Duncker was amazed by this poem, which he found quoted in Davenport's messy paper, but he was uncertain whether it genuinely documented the importation of canaries into Britain or whether these particular canaries merely symbolised the unwelcome Huguenots.

Islands of Canary Culture

As canary keeping grew more popular in the late 1700s informal societies began to emerge, providing an opportunity for men to socialise, exchange ideas and use their birds in competitions. These local federations, which met in coffee houses and taverns, forced common standards on what was expected of canaries because this was the only way to compete.[5] Transport was poor and human communities were still extremely insular, with pronounced regional differences in dialects and customs. It was precisely this combination of isolation and local tradition that resulted in the evolution of different canary breeds. Cities were effectively islands, each with its own environment in which close-knit groups of breeders expressed their fancy about what they felt was desirable in a canary. Just as with natural selection, the course of canary artificial selection was dictated by what was available. And what was available in England was strongly influenced by the Huguenot émigrés, who favoured the very large, upright canaries sometimes referred to as Old Dutch – more than twice as large as the original little wild bird. These big-bodied birds provided much of the raw material for the English breeders, whose strong regional preferences rapidly gave rise to different breeds. The industrial cotton towns of north-west England considered size crucial and developed a giant of a canary – the Lancashire. Just over the Pennines to the east, breeders strove to produce a very tall, upright bird, the Yorkshire canary, which to win prizes had to be slim enough – so they said – to pass through a wedding ring. In East Anglia a richly coloured, almost orange John Bull of a bird was preferred – the Norwich canary. In the Low Countries posture was deemed important, resulting in the hunch-back Belgium Fancy. It was later appropriated by the Scots who produced a bird with an academic stoop, known as the Glasgow Don. In France birds with long frilly feathers were favoured.

All these variations in size, stature and feather quality completely bypassed the Germans who, with a nationalist fervour that would persist unabated for centuries, continued to ignore the canary's appearance in favour of its voice.

The major thrust of selective breeding among the English canary breeders at the start of the eighteenth century occurred as men were 'improving' all sorts of animals through artificial selection. People described themselves as dog fanciers, pigeon-fanciers and chicken fanciers, but not, as far as I am aware, as sheep or pig fanciers. Nonetheless, selective breeding among farm animals was all the rage. Stockbreeders dressed up their activity as a patriotic response to a national need for bigger, faster-growing beasts to provide more meat for a burgeoning human population, but in reality they were no differ-ent from canary breeders in being driven by status. In one important respect, however, they did differ: livestock breeders who could indulge in experimental breeding were, without exception, very rich landowners with both the time and resources to breed bigger and better animals. Their wealth allowed them to work with huge numbers of animals, on a scale unimagined by most canary breeders, enabling them to make very rapid progress. Some of them were also very skilled. The stockbreeder Sir John Sebright boasted that he could produce pigeons of any type of feather in three years and any body form in six.

One might also think that there was a commercial incentive for livestock breeders but, paradoxically, ordinary farmers were extremely wary about adopting the new breeds of cattle, sheep and pigs. If wealthy breeders had gone to all this trouble to produce giant meat-making machines, why were farmers reluctant to take advantage of them? The cattle breeders' obsession with their own prestige drove them to produce beasts so extreme in size and shape that they were almost useless to ordinary farmers. The livestock breeders' self-indulgent endeavours resulted in huge-bodied, tiny-brained cattle that are only remembered now from their naive portraits.

The eighteenth and nineteenth centuries saw an explosion in the breeding of animal varieties.[6] In 1800 there were just fifteen recognised breeds of dog, but a century of artificial selection created no fewer than sixty varieties. The same occurred with virtually all forms of livestock, including the domestic fowl, the pigeon and the canary. According to the Soho cage maker Thomas Andrewes, when he published *A Bird Keeper's Guide and Companion* in 1830, the number of distinct canary varieties had increased from half a dozen to twenty.[7] He gave the most popular as the Lizard, Norwich, Yorkshire, Belgian and German, as well as the London Fancy. This breed still has a special place in the hearts and minds of canary breeders, and contemporary illustrations show it to be a beautiful bird; deep orange-yellow with jet-black wings and tail produced by a combination of genes that has since been lost – apparently for ever, since this breed became extinct during World War I.

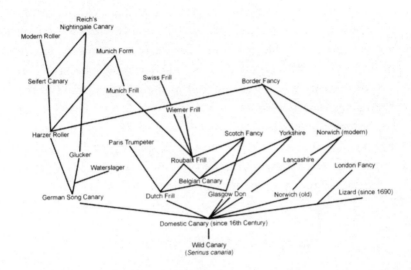

FIGURE 4 *Duncker's evolutionary tree for canaries, starting with the original wild bird. Notice Reich's canary positioned at the top left.*

When Hans Duncker finally completed his evolutionary tree for canaries (Figure 4) he knew it wasn't perfect. But given the uncertainties in the material he had at his disposal, it was the best he could do. Had they been alive, Hans Ehlers, his old supervisor, and Ernst Haeckel certainly would have approved, for this exercise would have been close to their hearts and no one since has done better. Nonetheless, Duncker must have wondered why Darwin hadn't done this for canaries, as he had for pigeons.

Duncker and Reich created this evolutionary tree together and Reich must have been very proud to see his nightingale-canaries on the uppermost branch of the tree. Their superior position was the result of generations of careful selection, as individuals with the best attributes were encouraged to proliferate. We have no idea whether Duncker and Reich ever discussed it, but the parallels with humans, and the German people in particular, were obvious. The similarities stemmed from a particularly persuasive form of social Darwinism that had begun in the 1870s. The famous biologist Ernst Haeckel pointed out that it was the Germans who 'deviated furthest from the common form of apelike men' and in whom one finds the 'symmetry of all parts . . . which we call the type of perfect human beauty'. This sort of thinking was still widespread in the collective mind of the German nation in the 1920s and as later events showed Duncker was enthusiastic about science improving society. We know nothing of Reich's views, but it would have been surprising indeed if the topic of improving society through careful breeding had not cropped up at some stage in their bird room conversations.[8]

Rapid Sports and Slow Selection

It was almost inevitable that Hans Duncker would turn next to consider the processes by which the different canary breeds had been produced. How had the early breeders created the different strains? Reich had no idea; all he knew was how fanciers like himself attempted to improve their existing stock. Fanciers made a conscious effort to breed only from the birds whose traits they admired, often pairing close relatives and rejecting individuals that didn't have the correct attributes. Intrigued, Duncker went back and reread Darwin, who gave him the more scientific answer he was hoping for. In the *Origin of Species* he said that in their quest for champions, pigeon-fanciers had employed two types of artificial selection. 'Methodical selection' – which was what all livestock breeders practised – consisted of breeding individuals together with the explicit goal of developing a particular trait such as body size or establishing an unusual 'sport'. 'Unconscious selection' was more subtle, but to Darwin much more important. It resulted from the cumulative actions of thousands of individual fanciers whose only objective was to breed together their best birds – but with no specific goal in mind. Generation after generation of unconscious selection resulted in changes that were virtually imperceptible in the short term, but dramatic when viewed over a longer period. To his cousin, William Darwin Fox, Darwin wrote, 'The copious old literature, by which I can trace the gradual changes in the breeds of pigeons has been extraordinarily useful to me.' The pigeon-fanciers' unconscious selection was Darwin's apposite and accessible analogy for natural selection.

In the three years prior to publishing the *Origin of Species* in 1859, Darwin had become obsessed with pigeons and had set up a loft in the garden at Down House. Encouraged by a friend, the eminent naturalist William Yarrell, he acquired numerous different breeds from a top pigeon-fancier in London. Darwin and his sixteen-year-old daughter Henrietta were enthralled by their new acquisitions and, writing to his cousin, Charles said, 'They are a decided amusement to me, and a delight to H[enrietta].' Darwin eventually had as many as ninety pigeons – including tumblers, pouts and barbs – in the Down House garden and it is a measure of how seriously he regarded his pigeons that after discovering Henrietta's cat had killed some of the birds he got Parslow, the family's long-standing retainer, to shoot it. Henrietta never forgave her father.[9]

Darwin was fascinated by all aspects of his charges; he watched their courtship rituals on the lawn and, together with Parslow, he dissected and skeletonised them in the potting shed. He joined two of London's most prestigious pigeon clubs; the Columbarian and the Philoperisteron. After several evenings of pigeon chat in smoky gin parlours, Darwin confided to his son that most pigeon-fanciers seemed to be funny little men, obsessed with producing and maintaining particular breeds. The 'almond tumbler' was their favourite, and despite his cynical comments Darwin was seduced: 'These are marvellous birds, and the glory and pride of many fanciers.' No fewer than 150 distinct breeds of domestic pigeon were recognised in the 1850s and although the fanciers themselves couldn't care less, they assumed each to be a separate species. In contrast, naturalists like Yarrell and Darwin were confident they had all arisen from a single wild ancestor – albeit guided by the grubby hand of man.[10]

Had Darwin ventured into canary clubs instead of pigeon clubs he would have encountered exactly the same obsessive desire to produce the best birds. Indeed, he would have found the same attitude in any of the numerous animal-breeding societies then in vogue in Victorian England.

In his quest to understand the canary's evolution, Duncker also consulted *The Variation of Animals and Plants under Domestication* (1868), Darwin's more detailed and distinctly turgid follow-up to the *Origin of Species*. As he turned the pages, he began to realise that generations of canary breeders, just like pigeon-fanciers, had been carefully selecting their birds. The only difference was that they hadn't been doing it for quite so long. Pigeons have been domesticated for thousands of years – hence the greater number of varieties. Darwin's concept of methodical selection of sports – selectively breeding from canaries with flecks of yellow plumage, or those that were all white – was the most obvious form of artificial selection and the easiest to come to grips with. Hans Duncker now focused on how this selection might have happened.

Animals with atypical colouring have always fascinated people, and museums are stuffed full of them, deluding schoolchildren into thinking that they are much more common than they really are. I remember as a child visiting my uncle's and aunt's farm in Norfolk and being shown a wild blackbird that, instead of being black, was the colour of a golden retriever. People with no other interest in birds came from miles around just to gawp at it. And it *was* spectacular – in forty-odd years of watching birds I have seen nothing like it since. Albinos or even partial albinos are among the commonest plumage aberrations and blackbirds with white patches are far from unusual in European gardens. Plumage aberrations like these sometimes result from disease. They can also be due to genetic defects induced by radiation.[11] It is now common knowledge that X-rays disrupt the genetic instruction manual, but in the 1920s it had only just become clear that radiation caused permanent genetic damage. Duncker had to point this out to his colleague Hans Meyer, director of radiography services for the Bremen hospitals, who was using radium, a potent source of gamma-rays, to treat cervical cancer in women.[12] One hopes that Duncker's warning was heeded, for by carrying the radium around with them in a little tin in their trouser pockets, the gynaecologists irradiated their own vital parts as well as those of their patients.

Mutations also occur by chance, albeit rarely, and their bearers, like the Norfolk golden blackbird, rarely survive long in the wild. But they can be nurtured in captivity and their genetic attributes captured and passed on to others.

Sports were perpetuated by breeding them back to their mother or father: a monumentally important trick technically known as a back-cross.[13] Most professional animal breeders I spoke to about this assumed that the back-cross was first used *after* the rediscovery of Mendel's work, but in fact plant breeders were using it in the 1750s and animal breeders even earlier. English racehorse breeders developed the technique in the late seventeenth century – through necessity. At that time they were importing Arab stallions, but not Arab mares, and had little choice but to mate their glamorous studs with less than glamorous home-grown mares. They then mated their cross-bred female offspring back to the original stallion, 'grading-up' generation after generation in a sequence of back-crosses designed deliberately to increase the proportion of Arab blood. By the early 1700s well-informed farmers were using the same technique to capture traits in sheep and cattle, and it is hardly surprising that grading-up was well known among administrators like Baron von Pernau, who in 1700 described the same method to produce a strain of white nightingales. Extraordinarily prescient in all his bird observations, Pernau also makes it crystal clear that the back-cross technique was well known to canary breeders at that time, indicating that this was precisely how they ensnared those first elusive patches of yellow plumage.

Once several canaries with yellow feathers existed, the change from green to yellow could in principle have proceeded fairly rapidly. However, back-crossing mothers with sons, or fathers with daughters, is the most severe form of inbreeding possible and inbreeding was a very mixed blessing. On the one hand it allowed breeders to capture traits very easily and even appeared spontaneously to create 'sports'. But inbred individuals were much more prone to disease and birth defects. Pairing closely related birds together resulted in more infertile or unhatched eggs, more hatching failures and more premature deaths. There seems little doubt that the canaries being bred in Germany in the 1500s and 1600s were subject to high levels of inbreeding, not least because, as all animal breeders know, some couples are more productive than others and there is often little choice but to continue to breed from these individuals.

Fanciers gradually began to realise that breeding in and in – what they also called consanguineous mating, literally the 'same blood' – was reducing the reproductive success of their stock. They also recognised that such problems could easily be rectified by out-breeding, that is by introducing new blood. But outbreeding was a dangerous game too because at one fell swoop it diluted, or at worst eliminated, all those traits the breeder had so carefully cultivated. A compromise was needed and, rather like royal dynasties, bird breeders resorted to something called line-breeding. They banned brother-sister marriages but encouraged matings between cousins to retain accumulated genetic or monetary wealth within families and used the occasional infusion of new blood to avoid the worst excesses of inbreeding, much as Anne of Bavaria was imported from abroad to inject new life into the French royal stock.

But sometimes even matings between cousins, like that of Darwin and his wife Emma, were too close, as Darwin began to realise, watching his brood of sickly children grow up. Illness and death permeated the Darwin household and in June 1850 Darwin's eldest daughter, nine-year-old Annie – his favourite – became sick with an illness that would not go away.[14] Her parents tried everything, starting with conventional medicine, such as it was, but when that failed they tried the new idea of sea bathing. And when that too failed they sent her to

Dr Gully's hydrotherapy establishment for the water cure – wet towels and cold compresses – which Charles had found so helpful with his own illness. But the water cure didn't work either and her desperate parents now resorted to folk medicine – a caged bird. Throughout Europe it was widely believed that small birds could cure a sick child by absorbing its sickness, and red birds like the crossbill, robin and bullfinch were thought to be the best – especially for scarlet fever – for the doctrine of signatures, common among herbal remedies, applied to birds too.[15] The link between redness and fever was obvious but, in fact, almost any kind of small bird would do and in October 1850 Charles and Emma bought Annie a yellow canary. She was enchanted by it. But as she herself recorded, within a few months the bird sickened and died. And by June the next year Annie too was dead – of consumption. Her parents were devastated and the pain of her passing lingered unabated for years. The canary had failed, just as it had done for Dickens's Little Nell ten years earlier.[16] When Darwin came to write his book on *Variation* seventeen years later, his memories of Annie were still raw and he could barely bring himself to confront the canary, glossing over it in a single page and making no effort to reconstruct its history as he had for pigeons. In fact, it was worse than this. Darwin, normally so meticulous, got his canaries muddled up and, in his haste to be done with them, referred to a feather-footed variety. There are feather-footed fowl and feathered-footed pigeons, but no such canary.[17] The canary, it seems, was Darwin's *bête noire*.

Broken Glass

At the time of Annie Darwin's death the canary was just beginning its meteoric rise in popularity. Between the time Mendel died, in 1884, and the recognition of his ground-breaking work in 1900, canaries were well and truly in vogue across Europe and North America.[18] It wasn't surprising that many biologists used them as model organisms to test

Mendel's ideas. Canaries were ideal research animals. They occurred in various forms: yellow and green plumage, crested or uncrested. They were prolific with a relatively short generation time and, thanks to the long tradition of canary husbandry, were very easy to rear.

It was these breeding experiments on canaries and other organisms that Duncker now focused his attention on. Most of these studies had been conducted in the previous twenty years and they encapsulated almost the entire development of genetics as a discipline. They also allowed Duncker to witness the controversies and excitement, the dirty dealing and euphoria as genetics struggled to emerge as a new discipline at the forefront of biology.

Mendel's findings had been simultaneously and independently rediscovered by three botanists in the course of their own studies of inheritance: Hugo de Vries in the Netherlands, Carl Correns in Bavaria and Erich von Tschermak in Austria.[19] But it was the Cambridge zoologist William Bateson who made it his life's goal to promote Mendel's work. Bateson read Mendel's paper on his way to give a lecture in London on 8 May 1900 and was sufficiently inspired by it to rewrite his talk on the train, and spent the rest of his life expounding Mendel's ideas. Bateson was a Yorkshireman, born in the old whaling town of Whitby and educated at Rugby, where he was described as a 'vague and aimless boy', and at Cambridge where he was thought arrogant and untidy. A colleague once said of him, 'William wasn't a person you were fond of, you admired him.'[20] Inspired by de Vries's work, Bateson had Mendel's paper translated and in 1902 published it together with his own *Mendel's Principles of Heredity: A Defence*. The defence was necessary because Mendel's revolutionary ideas were by no means universally accepted. The opposition consisted of a group of evolutionary biologists headed by Karl Pearson, a brilliant mathematician and statistician. The biometricians, as they were called, were sceptical about Mendel's results and claimed that Bateson and his followers were uncritical in their acceptance of his ideas. Both groups considered themselves evolutionary biologists and both were inspired by the writings of Francis Galton, but differed fundamentally in outlook.

Galton and Darwin were cousins, and Galton considered the *Origin of Species* to be among the most inspiring books ever written, but he could not accept the notion that natural selection worked on the tiny, almost imperceptible variations between individuals. Galton preferred the idea that selection operated on major 'sports' or discontinuous variation, and that evolution proceeded by discrete leaps. He wasn't alone; T. H. Huxley, Darwin's Bulldog and greatest ally, felt much the same. Bateson was convinced that evolution occurred as a result of fairly major shifts in characters and one of his objectives was to show that these shifts occurred in animals as well as plants. To this end he set up breeding experiments with poultry, mice and canaries at his home at Grantchester, near Cambridge, in the early 1900s. His basic goal was the same as Mendel's: to be able to predict the outcome of breeding together two individuals that differed in some clear-cut way. Bateson's canary work was carried out by Florence Durham, his middle-aged sister-in-law, whom he employed as a research assistant and who, unlike her shy and self-deprecating sister Beatrice, was a formidable and forthright woman.

The biometricians were Darwinian disciples, believing resolutely in natural selection as a gradual and imperceptible process, and were completely convinced that the tiny differences between individuals were perfectly adequate for its operation. They described their opponents' views as 'evolution by jerks'. The biometricians' approach was one of careful measurement and statistical analysis, seeking correlations such as that between the height of fathers and sons as evidence for heredity. Karl Pearson had stepped gingerly and somewhat paradoxically into the shoes of Francis Galton to become Professor of Eugenics at University College London in 1911. His brilliant collaborator was Walter Frank Raphael Weldon (known to his friends as Raphael), who later showed that the results Mendel got with his peas were almost too good to be true. It wasn't peas, however, but a nondescript little daisy that drove the first wedge between Bateson and the biometricians.

It started on 28 February 1895, when the botanist William
Thistleton-Dyer displayed two varieties of a flower called *Cineraria* at
a meeting of the Royal Society in London. One was the wild type *C.
cruenta* from the Canary Islands, and the other a cultivated variety
from the botanical gardens at Kew. The two forms differed dramati-
cally in the shape and colour of their flowers, but Thistleton-Dyer's
point was that the cultivated form had arisen from the wild type as a
result of artificial selection 'by the gradual accumulation of small variations'. Bateson wasn't at the meeting, but saw this as a chance to
challenge the prevailing Darwinian view of evolution. He dashed off
a letter to the premier scientific journal *Nature*, claiming that since
the cultivated plant was a hybrid of several different species, their
original offspring would have been extremely variable and it was from
some of these 'sports' that the cultivated form had been derived.
Thistleton-Dyer rejected the hybrid idea, but Bateson wouldn't back
down and a bitter controversy flourished in a stream of letters
published over the next two months. It was at this point that Raphael
Weldon stepped in to support Thistleton-Dyer, pointing out that
Bateson had not been entirely honest: '. . . and that the documents
relied upon by Mr Bateson do not demonstrate the correctness of his
views; and that his emphatic statements are simply a want of care in
consulting and quoting the authorities referred to'.

Bateson was incensed and took it as a personal attack. Nonetheless,
like true gentlemen, he and Weldon agreed to meet on 21 May to
thrash out their differences. The meeting was disastrous and the two
men never spoke another civil word to each other.[21] The rift between
Bateson and the biometricians grew wider with every new publication by the two sides, and wider still with the rediscovery of Mendel's
work. Pearson and his colleagues criticised their opponents for seeing
Mendelian ratios wherever they looked and for failing to consider any
alternative explanations, and accused the Mendelians of ignoring
results that were inconsistent with their views. For their part, the
Mendelians scoffed at their opponents' use of correlations and other

esoteric mathematical jiggery-pokery to decide whether traits were heritable or not.

The battle moved beyond England's shores to America, where Charles Davenport vigorously waved the Mendelian flag. A traditional zoologist, Davenport began his career studying the development of embryos, an interest that culminated in the publication of the encyclo-paedic and much-fêted two-volume *Experimental Morphology* in the 1890s. Soon after, he switched to study variation and evolution, employ-ing – at least for the time being – Karl Pearson's new statistical methods to analyse his results. Then, in 1902, after reading Bateson's translation of Mendel's famous paper, he underwent a conversion and became an ardent apostle of Mendel, proposing that the newly formed Carnegie Institute of Washington spend $10 million to establish a 'Station for Experimental Evolution' at Cold Spring Harbor, New York, with himself at its head. Two years later this dream had become a reality and Davenport, like Bateson in England, began exploring the inheritance of eye and hair colour in humans, and plumage traits in chickens and canaries. He was phenomenally productive and published over 400 articles – an extraordinary output even by today's highly competitive standards and something few scientists ever achieve.[22] But output on this scale rarely occurs without cutting at least a few corners.

In 1904 Davenport started to breed canaries to establish the patterns of inheritance in a semi-domesticated species. Earlier studies of hered-ity had been criticised for using species like chickens, which had been domesticated for thousands of years, and which some said were inap-propriate models for the study of inheritance in nature. The aim of Davenport's canary study was to address the questions: 'How is the crest . . . [and] . . . the plumage colour inherited?'[23] The investigation was based on an original stock of 'four yellow canary hens (one crested) of the short or German (Harz Mountain) type and two green birds of the same type. Also three yellow cocks (one crested) and two greens (one crested)' obtained from a New York dealer. Three years later Davenport was able to write, 'The interpretation of the results of breeding plumage colour is not difficult and may easily be brought to accord with Mendel's law.' His results with crests were rather less clear-cut, but Davenport nonetheless concluded that they behaved 'in a Mendelian fashion'. Even in a species that had been domesticated for as few generations as a canary, Mendel's rules still applied. Davenport also used his results to show that the 'yellow canary is derived from the original "green" canary by the loss of black' and rounded off his paper by writing, 'The plumage of a yellow canary may be compared with a letter that has been written with invisible ink. Wherever the developer acts (i.e., the black pigment of the green canary is added) that which is written appears with all of its idiosyncrasies.' When Davenport published these results in 1908, William Bateson immediately included them in his book, *Mendel's Principles of Heredity*, delighted to have more Mendelian ammunition.

Davenport's paper also caught the eye of a celebrated Scottish ophthalmologist and enthusiastic amateur canary breeder, Dr Rudolf Galloway. Unlike Bateson, Galloway was distinctly unimpressed and intensely irritated by Davenport's results. Indeed, they were catalytic in rousing this normally mild-mannered Scot to angry indignation, stimulating him to sort and analyse his own canary-breeding records accumulated over the previous eighteen years, to counter Davenport's claims.

Galloway was in a strong position to criticise Davenport because he knew more about canary genetics than just about anyone else at the time, and certainly more than Davenport. There was probably another reason why he felt especially confident in these matters. In 1908 he had produced the ultimate bird – a hybrid between a European siskin and a canary which was a clear yellow colour. Of the 526 canary x finch hybrids Galloway had produced during the previous seventeen years, this was the only one of its kind: the bird breeder's El Dorado. It was the most highly rated exhibit at the major Scottish shows in 1908 and the following year it was awarded first prize at the national bird show at the Crystal Palace in London.

In a block-buster article in *Biometrika*, a journal edited by none other than Karl Pearson, Galloway presented the results from his own canary studies,[24] summarily dismissing Davenport for not properly defining his terms, for not even knowing what breed of canaries he had used and, most of all, for being a sloppy experimentalist.

Davenport couldn't bear criticism and immediately drafted a rebuttal. The language of his letter was civil, but one can almost feel him composing it through gritted teeth. His anger, as we shall see, was fuelled by more than a twinge of guilt. Apart from a single reference to Galloway being 'scientifically untrained', Davenport's response is a model of self-restraint and ends on a conciliatory note, commending Galloway on his valuable results. But his attempt to appease the Scot fell on stony ground and Galloway retaliated in print with a further cutting comment: 'From the scientific standpoint only, his statements might be more consistent, though in all probability valueless . . . From his false assumptions . . . it is quite impossible for him to arrive at any scientifically correct, or practically useful conclusion.' One could hardly imagine a more devastating denial of Davenport's work, but immediately following Galloway's two articles in *Biometrika* there came another, much worse, this time from a David Heron.

Whereas Galloway had used the subtlety and precision of a surgeon's scalpel to expose Davenport's errors, Heron went in wielding a double-headed axe. Davenport must have known he had done a sloppy job, hence the rage, the guilt and the pathetic attempt to appease Galloway. Scientists can be cruel in their criticism of careless colleagues and scientists who do poor research lay themselves wide open for vicious scrutiny and opprobrium. Davenport's original twenty-one-page article provided Heron with virtually limitless evidence of incompetence but, perhaps for effect, Heron confined his comments to just 114 lines of text and four tables regarding the inheritance of crests. In this small part alone he identified no fewer than 109 errors. The same individual birds, for example, appear in one table as crested and in another as lacking a crest. Experiments were incorrectly labelled and Davenport ignored results that contradicted his Mendelian expectations, and he drew false conclusions from others. 'It can thus be shown . . .' wrote Heron, absolutely incredulous at the scale of sheer ineptitude he had discovered, 'that every conclusion made by Davenport can be proved to be false from a study of his own material; that if a fact has to be stated twice the one statement is flatly opposed to the other and that blunder is heaped on blunder until patience is exhausted. Yet such work is accepted as showing that Mendelian rules apply to Canaries!' – by which he meant the inclusion of Davenport's results in Bateson's book. Few scientists ever receive a public flogging like this and it demonstrates the strength of feeling and the desperate quest for supremacy that motivated the warring tribes.

Who was this David Heron who so savagely attacked Charles Davenport? None other than Pearson's research assistant. Heron's hatchet job on Davenport was merely Pearson's way of opportunistically reprimanding a turncoat and part of his ongoing campaign to discredit the Mendelians.[25]

Genetic Isolation

When Duncker began his own genetical studies in the early 1920s the dispute between the Mendelians and biometricians had been resolved. If they hadn't all been so bloody-minded they would have realised much earlier that there was no real disparity in the two visions of evolution. For fifteen years both sides had failed to see that their different approaches were little more than the complementary processes of 'methodical selection' and 'unconscious selection' that Darwin had highlighted fifty years previously. The crux of the issue was that some traits, like human blood groups and the characters that Mendel had chosen so carefully in his peas, were discontinuous – one thing or the other: A or B, smooth or wrinkled, green or yellow – and controlled by single genes. Other traits, meanwhile, such as human stature, varied continuously and were, crucially, controlled by multiple genes. Regardless of whether the traits were discontinuous or continuous, Mendel's rules still applied – albeit in more subtle ways than the early geneticists ever imagined.

The papers by Davenport, Galloway and Heron played a crucial role in Duncker's personal development as a geneticist and he cited their results in his own huge account on canary variegation. Making sense of Davenport's paper wasn't easy and, like Heron, Duncker had to go through it with a fine-tooth comb. Clearly much of what Heron and Galloway had said about Davenport was true, but fortunately, amidst the mess of errors, Davenport had included all his breeding results in a table of raw data, enabling the endlessly patient Duncker to sort fact from fantasy. Davenport finished his study by saying that there might be one or two 'factors' – what we now call genes – controlling variegation in canaries, a conclusion that broadly coincided with Duncker's own, and confirmed that canary colours were inherited in a Mendelian manner.

The fact that Duncker should choose to study heredity and to a large extent succeed is absolutely remarkable. For one thing, he was a middle-aged schoolteacher rather than a practising research biologist. For another, in intellectual terms Duncker was completely on his own. Germany had no tradition or interest in studying the mechanisms of inheritance, and as the great historian of biology Ernst Mayr said of this period, 'The average German biologists were unquestionably backward in their understanding of genetics.'[26] But Duncker was anything but average, either in intellect or ambition, and his work was exceptional in the history of genetics in Germany.

It seems strange that the Germans should have been so backward in the study of inheritance when, in terms of evolution, they were said to be more Darwinian than Darwin himself. This was thanks largely to biologist Ernst Haeckel who, after reading the *Origin of Species* and having what he felt was a near-religious experience of meeting Darwin at his home in England in 1866, became utterly convinced of the truth of evolution and proceeded to Darwinise Germany through his popular science writings. But Haeckel's views on the mechanism of evolution – the actual manner in which traits were transmitted from parents to offspring – remained hopelessly Lamarckian and he was preoccupied with 'improvement'. Haeckel muddied the waters, in more ways than one. It wasn't that German biologists had no interest in the mechanisms of evolution, they had, and ever since the 1870s had been obsessed with the genetic blueprints that allowed an egg to give rise to a bird, or an acorn to an oak. But they were too preoccupied with animal development and structure to worry about the nuts and bolts of heredity itself. In much the same way, the behavioural ecologists seventy years later studied the evolution of behaviours which they assumed were inherited but without making much effort to check.

If there was little interest in inheritance inside German academia the situation elsewhere couldn't have been more different. In England, Bateson's work had generated a tidal wave of studies aimed at elucidating the secrets of inheritance, particularly of traits like coat colour in rodents. In the USA Thomas Hunt Morgan, fascinated by development and mutations, chanced upon a white-eyed mutant among his fruit fly stocks in 1907, luring him irrevocably into a pioneering and revolutionary study of transmission genetics. The next twenty-five years saw Morgan's fast-breeding fruit flies relinquish their hereditary secrets at an ever increasing rate (and much more quickly than Bateson's chickens, mice and canaries), providing the foundations for modern genetics.

Despite the dramatic progress elsewhere, there were still no chairs in genetics at German universities in 1921 when Duncker and Reich started their studies. The reason lay partly in Germany's antiquated and hierarchical academic system in which students and younger researchers dare not challenge their professors' views; a tradition that disrupted the influx of new genetic knowledge from the United States and Britain. At this time zoology teaching in German universities was

Hans Duncker aged 31; young and handsome before his canary studies.

A German canary breeder's birdroom in the early 1900s. I imagine Reich's birdroom to have looked much like this.

Hans Duncker and Karl Reich some time in the mid-late 1920s.

General consul Carl Cremer in the 1930s: suave, smooth and confident.

Jan Steen's *Rustic Love* – also known as 'Couple Flirting Outdoors'.

Brueghel's *The Return of the Herd – Autumn, 1565.*

Lancret's *Spring*, a symbolic bird-catching scene. This image was used by MacPherson in his book on bird catching to illustrate the link with sexual selection.

Chardin's *The Bird Organ* or *A Woman Varying Her Pleasures*.

The original wild canary.

The domesticated yellow canary.

The plate from Duncker's paper (Duncker 1927b) showing a male red siskin, the dominant white canary and their disappointing offspring.

The red canary.

done entirely by a professor and one or two assistants. Salaries were poor and professors supplemented their income with the fees from extra teaching, so they naturally chose to teach the large comprehensive courses that brought in the most income. The younger staff, who were most aware of recent developments, taught the smaller, more special-ised courses. So genetics either didn't get taught or, if it did, it played second fiddle to developmental embryology.[27] The almost total intel-lectual isolation in which Duncker worked makes his success all the more remarkable, although perhaps it was precisely because he lived and worked outside the academic system that he achieved so much.

In late 1924, having successfully completed the studies of variega-tion and confident that he understood the inheritance of canary colours, Duncker adopted an utterly audacious plan. It was an idea that flew in the face of everything he had learned about the single-species origin of domestic breeds. His idea was to combine the genetic material of two distinct species to create a new one – a transgenic canary. This was the first attempt to create a genetically modified animal, the first bit of animal gene technology, and once again Duncker was decades ahead of everyone else. The quest for the red canary had begun.

7

Mixed Blessings

I am enclosing you a very rough water-colour sketch of a lovely little bird that I bought in Santa Cruz [Teneriffe], and whose proper name I do not know. Can you identify him for me? I have never seen him in England, which may be an oversight on my part. It is a finch from Caracas (South America) which is imported to Teneriffe, and mated with the Wild Canary, producing a mule-bird like a Wild Canary dipped in saffron.

Letter from the REVEREND HUBERT D. ASTLEY to R. PHILLIPPS (1902)

By the 1870s songbirds could be heard and seen in virtually every home in Europe. In Britain, Queen Victoria was still moping – despite the invention of the telephone and the best efforts of Gilbert and Sullivan – over the loss of her Albert, now ten years dead. At Down House Darwin was alternately busy and sick, pursuing the glory of his biological intuition. In France the Impressionists were using colour in ways never seen before and in Germany Otto von Bismarck presided over a newly unified country – the Second Reich.

The British National Cagebird Show of 1873 was held in Norwich at the Assembly Rooms. A popular event since its inception in the 1850s, this one attracted a huge crowd. The main hall was full of people squeezing and pushing between the long lines of trestle tables stacked with cages, desperate to catch a glimpse of the exhibited birds. Amazingly, despite the thick fog of cigar and pipe smoke that filled the hall, the voices of the birds rose cheerfully above the hubbub of the crowd. But the exhibition hall was buzzing with more than the sound of birds. The buzz was a human one and centred on a sensational set of Norwich canaries exhibited by a Mr Edward Bemrose from Derby.[1] They were a bright, luminous orange and swept the boards, taking 'Best in Show'. The Norwich canary – the Victorians' favourite breed – was judged partly on its shape and posture but primarily on its colour. Rather than being a plain lemon or even gamboge, to win the Norwich had to be a rich yellow tending towards orange. Bemrose's birds, the colour of marigolds, were more vivid than any previously shown and few in the Assembly Rooms on that day could quite believe their eyes. Rumours were rife. Had Bemrose fed his birds something exotic, or had he plunged them into dye to produce such a winning colour?

Orange canaries had been seen at shows in the previous two years, and one judge described them as 'bursting upon the bird world like meteors'. Their owners claimed to have bred them rather than fed them, but many in the fancy were deeply sceptical – for two reasons. First, the men walking away with the prizes at these shows had produced entire broods of orange winners, whereas normally only one bird in a hundred ever made a champion. Second, fanciers who had bought these beautifully coloured birds – often for considerable sums – were later devastated to see them fade into jaundiced mediocrity after the autumn moult. At one show the judges broke all the rules by pulling a bird from its cage and attempting to rub off what they assumed to be applied colour. But all the rubbing did was to destroy the starlet's tail feathers.

The first sign that things were starting to turn nasty occurred at the Crystal Palace show in February 1873, when Bemrose, described by W. A. Blakston, a top canary judge and author, as 'one of the keenest fanciers of the day, and a man on whom no one could lay a finger of suspicion', exhibited two wonderfully pigmented orange birds.

Despite his impeccable credentials, Bemrose left the show under a cloud of suspicion and vowed defiantly that he would be back with even more winners. True to his word, at the new season in October the same year, Bemrose took every prize within reach at the great Norwich Show. His unsuccessful opponents, the Norwich canary breeders, were enraged and eventually rose up as a body to challenge him. Seven of Bemrose's winning team suffered the indignity of having their orange feathers plucked and sent off for analysis, but the County Analyst's Office found nothing irregular on the birds' plumage. This was true, for the birds had certainly not been dyed – at least not externally. But by the end of the year Bemrose's colourful secret had leaked out when one of his confidants foolishly *sold* the knowledge to another fancier for £50. Such behaviour could not be tolerated. Canary breeders were gentlemen and were expected to behave accordingly. Artificially colouring a canary was one thing but selling a secret for personal gain was a disgrace to the fancy. As soon as the society officials learned of this breach of etiquette they informed Bemrose, who then had little choice but to set the record straight by going public. He published his secret in the 11 December issue of the *Cottage Gardener*.[2] The magic process that had so demoralised the rest of the canary fancy turned out to be utterly trivial: red peppers. Fed to them during their moult, red pepper pods transformed yellow canaries into bright-orange ones.

It was inevitable that if points were given for a canary's colour, some men would stoop to devious means. Responding to the Bemrose scandal, George Barnesby,[3] another well-known judge, pronounced, 'The artificial colouring of birds, especially canaries, is a tricky artifice often practised by some, who much to their shame and disgrace, resort to the defacing of Nature's works for the sake of gain.' But they were all at it and prior to the discovery of red peppers fanciers had tried everything they could think of to enhance the intensity of their birds' colour: marigolds, fuchsias, nasturtiums, port wine, beetroot, saffron, mustard seed and cochineal. It was only because their previous potions had produced little change that their use was never an issue. Red pepper was different: its effect was so dramatic that it took the breeders' breath away.

Once the red pepper effect became public knowledge everyone started to use it and colour feeding, as it was called, became an accepted and legal part of the business of producing Norwich canaries. It had to: if Norwich fanciers were to compete on a level playing field, there was no other way. And they were optimistic. The future of the Norwich canary was bright and the future was orange. But not red. Feeding canaries peppers turned them the same hue as tangerines, peaches and marigolds, but however much red pepper fanciers forced their birds to eat, the canaries remained resolutely orange.

Reich must have told Duncker this story, which was a well-established part of canary folklore, during one of their many evening conversations. It probably made him laugh but it also did something else. It planted the seed of an idea in Duncker's brain.

There was yet another example of men changing the colours of birds' feathers that Reich probably knew about and related to Duncker. This was the trick of enhancing the colour of parrots by anointing them with frogs. The very notion sounds preposterous and it is easy to imagine Duncker dismissing Reich's story as apocryphal. But Reich assured him that since no less an authority than Comte de Buffon had described it it must be true. Buffon's description in his 1790s encyclopaedia of what he called 'the artificial parrot' probably came from 'Dom' Pernetty's account of Louis-Antoine de Bougainville's voyage of discovery to South America in the 1760s. Pernetty described one particular parrot thus:

> All its plumage, especially the head, neck, back and belly is sprinkled with feathers, some of them yellow like daffodil, or yellow like lemon, some others carmine red or crimson red, mixed with feathers either more or less dark green, or bright blue especially around the ears. This type of plumage is due in part to nature and in part to the

art. When the bird is very young, and has only got its feather-sheaths grown out after the down feathers, sheaths are plucked in several points, and instead is immediately inserted a sort of poison like liqueur. Feathers that grow after the sheaths then become yellow or red, instead of green as they should have been naturally.

Duncker and Reich almost certainly dismissed this as nonsense and of no possible relevance to their own work, but as unlikely as it seems, the change in colour in these parrots was not only genuine but genetic in nature. The 'liqueur' came from a frog – one of the colourful but highly toxic dart poison frogs, azure blue with gold stripes – aptly known as the dyeing frog *Dendrobates tinctorius*. Apart from transforming parrots, the most remarkable thing about this frog was something Duncker could hardly have anticipated: it could create genetic change in anything it came in contact with. The poisons in the frog's skin are alkaloids, protecting the frog from bacterial and fungal infections. The ability to produce these toxins is determined partly by the frog's genes and partly by what it eats. Frogs that in the wild could kill a man with one lick are rendered entirely harmless in captivity when deprived of their natural diet. With appropriate feeding, however, they rapidly regain their alkaloid secretions. Alkaloids are mutagenic – they cause genes to mutate. More familiar mutagens include X-rays, other kinds of radiation and ultraviolet light, but there are numerous chemical mutagens too, and many of them cause cancer. A growing feather is a mass of rapidly dividing cells and applying the frog's alkaloids was like bombarding the feather genes with mutagenic missiles. The aim was to hit only those genes that control colour, but the technique was so crude that the mutagens usually struck many other genes and utterly disrupted the bird's ability to produce proper feathers. Few birds survived the treatment. But in a handful of instances, just as Pernetty reports, the mutagens found their target – hitting only those genes controlling the biochemical pathways that determine the final colour of the feather and leaving everything else intact. These multicoloured parrots were much sought after and sold in the Paris bird markets for large sums.[4]

Buffon's artificial parrot was in a sense genetically modified, but since its creators had no idea they were tinkering with genes and because the parrot was unable to transmit its new genes to its offspring, it hardly counts. Nonetheless, had Duncker known that the effect was genetic he would certainly have been intrigued.

Perhaps on that same evening, Reich also told Duncker about a well-to-do Bremen bird-fancier, Consul Carl Cremer, who was keen to meet him. Cremer was a successful budgerigar breeder, president of the Austauschzentrale der Vogelliebhaber und Züchter Deutschlands[5] – the German bird-keeping society known as the AZ for short – and a member of the Bremen Natural History Society to which Duncker also belonged. Although Duncker knew him to nod to, they had never really spoken because thus far their interests in birds had been rather distinct. But following the publication of Duncker's eight-part account in the popular cage bird paper *Gefiederte Welt* (Feathered World) in 1924, describing his successful unravelling of the canary's tricky variegation helix, Cremer was anxious to talk to him. He wondered if Duncker's genetic talents might help him breed better-coloured budgerigars.

Consul Carl Hubert Cremer was a tall, imposing businessman with a goatee and an obsession for foreign birds and budgerigars in particular. He had made his fortune as a merchant and a shipbuilder. Well-travelled, well-educated and extremely well-heeled, Cremer thought nothing of spending the equivalent of £500 on a single bird of a new colour.[6] We don't know for sure, but it seems likely that Cremer survived the early 1920s recession through his foreign invest-ments, as other wealthy merchants had done.

The meeting took place at Cremer's house in the autumn of 1925. Duncker made the short walk from his own home across town to where Cremer lived at No. 130 Am Dobben, an exclusive street not far from Reich's hardware shop. Even before he rang the doorbell

Duncker could see that, with its neoclassical façade and four floors, the house was the grandest of a magnificent row. A maid showed Duncker into the drawing room where Cremer came to greet him. Duncker was immediately impressed with Cremer's friendly, forthright manner, by his dynamism and not least by his obvious prosperity. Duncker had never seen such affluence devoted to bird keeping. Cremer took Duncker on a guided tour of his aviaries. The first contained a dazzling collection of tropical jewels: orioles, tanagers, robins, thrushes and red siskins. In the next there were dozens of budgerigars, loud in voice and colour. Then on to Cremer's breeding rooms to see the bloodstock. Here there were budgerigars of various shades of green, sky-blue, cobalt, mauve, yellow and white. Finding that Duncker knew nothing about budgerigars, Cremer explained how in 1840 the original pair had been sent to Europe from their native Australia by John Coxen, the brother-in-law of the famous naturalist John Gould. Like the canary, the original wild birds were green – albeit a much brighter green than the canary. The yellow mutation was the first to appear, Cremer continued, in 1870, and blue birds appeared just eight years later. The main English bird keepers' handbook published in the 1870s had commented prophetically, 'Without doubt another ten or twenty years will witness as great results of intelligent breeding of varieties of the Budgerigar as has been witnessed in the case of the Canary.'[7]

Cremer then showed Duncker his most prized possessions, his grey-winged budgerigars, one of the most recently established and most highly valued of all mutations. To Duncker these birds didn't seem as attractive as the all-blue or all-yellow ones, but he knew from experience that beauty for a bird keeper was often in the eye of the beholder.

The two men retired to the palm-filled conservatory and continued to chat. From their comfortable cane chairs they looked out into the aviaries built on to the back of the house. Captivated, Duncker saw more than just birds; he also saw the enormous potential of

Cremer's facilities. The bird room in Reich's home where they had done the variegation experiments had been adequate, but it was very modest by comparison. Duncker's mind raced with the possibilities that Cremer's aviaries offered.

Cremer started to tell Duncker about himself; how he was the eldest of seven children and how his two brothers and four sisters had teased him by nicknaming him 'the principal' because of his obvious enthusiasm for inheriting the family distillery business. He had been useless at school and bored by what he laughingly called the dried talk of petrified teachers. At sixteen he left school and started in his father's business and at the same time began keeping and exhibiting budgerigars and canaries. He laughed again as his memory caught up with him: business and bird keeping kept him pretty busy, but not too busy, he said, to pursue the ladies. Seeing Duncker wince, Cremer quickly continued, describing how he had married the daughter of his father's business partner. After his marriage he started to travel, setting up businesses wherever he went, in Antwerp, in Paris (where he had a wine business), in London and Oporto, eventually returning to Bremen where he met his future colleague Kühlke with whom he established an export company in 1892. This venture did well, and he then started his own ship-fitting organisation, which was also success-ful. During World War I he joined the diplomatic service in the Netherlands and was made Generalkonsul. Cremer stopped: this was his story, he said.

Cremer's openness and enthusiasm encouraged Duncker to say what he had been thinking. Instead of simply keeping all these birds as decorations, why not use them to solve the great mystery of colour genetics – in budgerigars, canaries or, indeed, any species he wanted? Working as a team, Duncker said, they could do great things. Cremer beamed: he was hooked – it was the idea of budgerigar experiments that convinced him – and he happily agreed to provide whatever was needed. Better still, he told Duncker he was also happy to support his canary studies. But Cremer had another surprise in store. They wouldn't conduct their proposed studies here in town, but instead would use his country house, Rosenau Villa at Vahr, two or three kilometres away on the east side of Bremen, where there was a bit more space.

A few days later, Cremer took Duncker to see for himself. Named for its rose gardens, Rosenau was huge: the gardens extended over six hectares and were bursting with exotic trees and shrubs. There were ponds with fountains, stables for the horses and in one corner a well-ordered vegetable garden. The house was typical of the mansions so loved by Bremen's rich merchants: newly built of red brick, picked out with pale mortar and adorned with gracefully arched windows. In the grounds Cremer kept rare breeds of chickens, ducks and turkeys, and there were yet more aviaries full of exotic birds. Duncker laid it on the line to Cremer, saying that to understand the inheritance of colour, be it in budgerigars or canaries, it was absolutely essential to raise very large numbers of birds to obtain statistically valid estimates of the ratios of the different colour forms – just as Mendel had done with his peas. Cremer didn't need much persuading: he threw money at the project, and within a few weeks of meeting Duncker he had workmen busy building new two-storey bird houses with aviaries inside and out, and hundreds of cages to hold all the birds they would need for their experiments. Duncker's brain and Cremer's cash created a unique synergism. Karl Reich wasn't excluded; he was there to advise on the canary research. Almost overnight Rosenau meta-morphosed into a research institute in which Duncker, Reich and Cremer would change the face of bird genetics for ever.

Inevitably, the initial thrust of the research was budgerigars. They started by trying to understand the genetic basis for Cremer's grey-winged mutation. But Duncker's real passion still lay with canaries and an idea that Reich had unwittingly planted in his brain. It was seeing the beautiful red siskins in Cremer's aviary that had fertilised the latent red canary idea. Why not breed a red canary? For Reich, breeding was everything, just as it was with his nightingale-canaries

– if an effect couldn't be bred, it wasn't worth having. For the higher echelons of German bird keepers nature was vastly superior to nurture in every respect. It was the challenge that mattered; any clown could feed a bird red peppers to turn it orange, but it took genuine skill to figure out how to *breed* an orange bird. Still more to breed a red one. Ever since that conversation, the idea of creating a red canary had worked away at Duncker's neurones. The sight of the fiery red male siskin in Cremer's aviaries caused it to flutter free and into the light. This was exactly the type of challenge Hans Duncker loved: it tested his ingenuity and pitted him against nature. Here was a small finch carrying vivid red genes and he was going to put them inside a canary.

Cremer told Duncker the story of how the red siskin had made its appearance in European aviculture.[8] With its brilliant plumage and sparkling song the red siskin may have been kept as a cage bird in Venezuela even before the Spanish first appeared in 1530. With their long tradition of bird trapping and trading, the Spaniards appropriated the red siskin and transported it, along with dozens of other species, back home and to the Canary Islands. The two species crossed in the night as canaries were exported to South America and it was probably there that the first hybrids were created. For almost 300 years, however, the red siskin remained unknown to the rest of Europe – the Spanish at this time vigorously regulated the trade in wild canaries and kept a firm grip on their siskins too. It wasn't until 1820 that the red siskin became known to science – centuries after most other known birds had been officially described. It was William Swainson, an avid English ornithologist and artist who spent much of his time in the Mediterranean, who first recognised the red siskin as something new. He came across a single male in the possession of an Englishman who told Swainson he had obtained it on the 'Spanish Main'. In the following fifty years very few other red siskins were seen, until at the peak of the cage bird frenzy in the 1870s a few were imported into Continental Europe. August Wiener, co-author of one of the most magnificent English volumes on bird keeping, *The Book of Canaries and Cage Birds*, had acquired a single male in 1877, but didn't think the species worth including in his book because it was so rare. Nor did the red siskin appear in Robson's and Lewer's follow-up volume, *Canaries, Hybrids and British Birds in Cage and Aviary* published in 1911, possibly because unlike its predecessor this volume was explicitly about British birds. Even so, the red siskin might have had a mention since by this time several people kept them. More significantly, several fanciers had successfully crossed them with canaries.

Hearing that bird keepers had already hybridised red siskins and canaries, Duncker needed no more encouragement. The logic was straightforward: in all other crosses between a finch and a canary – which breeders referred to as 'mules', to distinguish them from hybrids between two finch species – the cross-bred offspring almost always assumed the colour of their finch parent rather than the canary parent. Since the red siskin possessed red genes, it followed that a hybrid produced from a yellow canary would be red rather than yellow. Duncker's mind raced ahead of itself: he and Reich would breed some red siskin mules; the following year they would breed the mules together and generate even redder offspring. By selecting only the reddest of their offspring for back-crossing with canaries, they would end up, in four or five years, with a pure red canary. It all seemed so simple, so straightforward and so exciting. Even Cremer was enthusiastic; budgies were one thing, but he readily acknowledged that a red canary would be utterly novel.

Mixed Encounters

The term 'mule' originally referred exclusively to the cross-bred offspring of a donkey and a horse but has since been applied to hybrid plants, hybrid fish and in the bird-keeping fancy to the offspring of a canary and a finch. Textile workers, many of whom were also canary-mule breeders, even referred to one of their machines as a 'mule' because it comprised a mixture of Joseph Arkwright's warp machine and Mrs Hargrave's hand jenny. 'Hybrid' is similar to mule in that it, too, once had a very specific meaning, identifying the offspring of a female domestic pig and wild boar, but it now refers to the result of cross-breeding any two species of plant or animal. The term hybrid comes from hubris – insolence against the gods – reflecting the ancient view that there was something improper about inter-specific crosses. German bird keepers used the term 'bastard' to describe both mules and hybrids and it regularly appears in Duncker's papers with reference to red siskin mules. Bastard was also used in Germany to describe people of mixed race.

Hans Duncker must have known that the terms 'mule' and 'sterility' invariably went hand in hand, and what was true for the mixed offspring of horses and donkeys was also true for other inter-specific crosses – including canaries and finches. Nonetheless, he hoped that they could somehow either leapfrog or ride roughshod over this general rule and generate *fertile* offspring from canaries crossed with red siskins. Duncker was confident that he could persuade red siskins to breed with his canaries and produce red mules. He had it on good authority that the Spaniards, both in South America and on the Canary Islands, had a long history of doing this. Indeed, ever since canaries had been bred in captivity, they had produced mules with other finches, mixtures that must have originally arisen by accident when female canaries and male finches were kept in the same aviary. One of the earliest examples was a beautiful goldfinch mule painted by Lazarus Röting in 1610. At that time only male finches were kept – for their song – and as the breeding season progressed these birds must have felt more and more like men in singles bars whose threshold for what is acceptable in a partner declines rapidly as closing time approaches. For a male finch, mating with a female canary was better than not mating at all.

When the breeders of mules and hybrids are successful they

produce what are effectively new organisms. Some bird keepers find this prospect irresistible, not least because many crosses are stunningly beautiful in both feather and voice. As soon as it became widely known that creating mules was a possibility, bird keepers set about trying to breed them. By the time Hervieux wrote his book in the early 1700s, mule breeding was extremely popular:

> It being natural for Man never to be satisfy'd with what he has, but to despise what is in his Power, and ardently to desire whatsoever is out of his Power; curious Persons at present act accordingly in respect of Canary-Birds. They are not satisfy'd with having an Abundance of Canary-Birds of the most beautiful sorts; but are for altering their Nature, and most of them take pains to make Canary-Birds couple with Birds of another sort, whose Young are call'd Mongrels.

Breeding mules has retained its hold over bird breeders since Hervieux's day and it is now among the most status-driven part of the fancy. The lure is the ability, through a combination of skill and luck, to triumph over nature. Success is rewarded on the show bench and the more difficult two species are to breed together, the greater the achievement. Mules and hybrids often win 'Best in Show' at national competitions today merely on the basis of their implausibility.[9]

Although accidental matings between different species sometimes occur in captivity, breeding mules and hybrids to order is more difficult, and breeding birds that will win exhibitions even more so. In 1877 the cage bird judge George Barnesby wrote,[10] 'If anything in bird breeding tests the patience of a true fancier most, it is mule breeding'. It is precisely because of this challenge that bird keepers remain attracted to this branch of the hobby.

Improbable Crosses

For the past 200 years mule and hybrid breeders have scored points over their fellow fanciers either by producing a hybrid from an unlikely combination of parent species, or by breeding a clear yellow mule – a cross between a canary and a finch with no dark feathers whatsoever, like Galloway's famous mule of 1908. Duncker's ambition was to achieve both these goals simultaneously, producing descendants of canary x red siskin mules with no dark feathers, but with the siskin's red ground colour rather than the canary's yellow.

Some mules are easy to produce. The goldfinch mule so lovingly illustrated by Lazarus Röting is a case in point: simply putting a male goldfinch into a cage with a female canary at the height of summer will invariably produce hybrid babies. The serin, the canary's closest cousin, is the best of all muling finches and even produces completely fertile offspring. But this is like crossing wolves and dogs. They are effectively the same species. Easy mules and hybrids were of little interest to bird-fanciers – there's precious little prestige in doing something simple.

Other mules are either too difficult or impossible. Johann Bechstein, the great German authority on cage birds, fantasised about a hybrid nightingale x canary; a bird, he said, that would be the ultimate songster. He and dozens of other fanciers in the 1700s and 1800s must have tried it, but with absolutely zero success.[11] Canaries that sang nightingale songs were the closest they ever got.

In his bird-keeping book, Eleazar Albin included a colour illustration of an extraordinary-looking bird that he said was a hybrid between a cock swallow and a hen goldfinch. The bird is indeed like no other and its inclusion in the book was a clever sales ploy: such an unlikely hybrid must have fired the imagination of every bird-fancier who saw it. Were it produced now, it would probably score highly on the show bench for its sheer novelty, but in reality Albin's extraordinary 'hybrid' was nothing more than an aberrant goldfinch whose striking blackness may have been either genetic or environmental in origin.[12]

Impossible hybrids were like fishermen's stories of ones that got away: they conferred no kudos. What was needed was something unbelievably rare but not totally impossible, and when Bechstein wrote his monumental handbook on cage birds in the 1790s, he described a hybrid of exactly this type.[13] It was a cross between a cock bullfinch and hen canary, bred by a fellow German. Mention of this extremely improbable hybrid (there was no picture of it) must have rattled the cages of bird keepers everywhere and spurred them to try to create one for themselves.

Who knows how many fanciers struggled, season after season, with a bullfinch cock and a canary hen to no avail? There may have been one or two false hopes – the occasional fertile egg and even the odd hatchling – but none of them survived. Eventually, a full century after Bechstein had first raised their hopes, a canary x bullfinch mule was reared to maturity – but contrary to the usual muling practice, this bird's father was a canary and its mother a bullfinch. And what a bird it was – such as England had never seen before – one of the most lusciously coloured hybrids ever produced. At the Crystal Palace exhibition of 1898 its owner, John Williams of Liverpool, shot to stardom and took first prize – the highest accolade an exhibitor could hope for. But no sooner had he won than the floodgates of controversy burst. Suddenly there was concern whether the bird really was what Williams claimed it to be. The judge, Charles Houlton, was convinced, but much of the rest of the fancy remained aggressively sceptical. There is much at stake in such contests and bird breeders are deeply suspicious of extravagant claims, even from breeders like Williams, whose breeding skills had earned him the title of 'Mule King'. Because a canary x bullfinch mule was so incredibly unlikely and because the bird's parentage was so ambiguous, many felt it totally unacceptable to award Williams first prize. From the bird's overall appearance one of its parents was clearly a bullfinch, but the other parent was far from obvious. The uncertainty was made worse because Williams had not bred the bird himself but had bought it from another fancier who, to confuse things further, swore that it was a cross between a linnet and a bullfinch. One of the most vociferous critics was Lambert Brown, whose greenfinch x canary mule had been placed second at the same show. Even after the National British Bird and Mule Club had agreed, following protracted discussion, that the bird really was a canary x bullfinch mule, Brown was still protesting. The issue of the mule's authenticity rumbled on for years. So much so that when Charles Houlton came to write his book *Cage Bird Hybrids* years later he was still worked up about the bird that had caused him so much anguish.[14] 'To write about this wonderful cross . . . my pulse beats and my heart literally throbs with its earnestness and fullness, so that I am almost too full of the subject to know where . . . to start.'

Since then there have been several canary x bullfinch mules and they have all been produced using a male canary and hen bullfinch. Plenty of fanciers have tried with bullfinch cocks but with absolutely no success. What renders the male bullfinch such a useless hybridiser? The answer is that he is sexually dysfunctional – hopeless at sex. The mind is willing but the cells are weak. One could accept that bullfinch cocks might be hopeless hybridisers if they were particularly picky about who they copulated with but, as breeders are quick to point out, they aren't. Rather, the reason lies in the fact that female bullfinches are incredibly faithful to their partners – unlike most other birds, whose promiscuity is notorious.[15]

The link between the female bullfinch's sexual fidelity and a male bullfinch's inability to sire hybrids is this. When females of a species routinely mate with several males (of their own species, of course), there is intense competition between the cocks to fertilise a female's eggs. The cock with the most effective sperm passes on his successful genes to his sons. Having efficacious sperm depends to a large extent on overcoming the female's reproductive defences in which she reduces the number of sperm from the millions the male gives her to the handful she actually needs to fertilise her eggs. When females are promiscuous the sperm numbers problem is intensified – especially since each male wants the female to retain *his* sperm at the expense of the other guys'. Males of those species with promiscuous females thus produce more sperm and, crucially, sperm that are much better at jumping over the female's ticket barrier and getting to her eggs. The poor male bullfinch hasn't got a chance. Aeons of female monogamy have blessed him with a derisory sperm count and a limp ejaculate. His sperm are sufficient for fertilising females of his own kind, but in the hostile and alien oviduct of another species they are utterly impotent.

Knowing the male bullfinch's deficiencies, Duncker prayed that the red siskin might make a more tractable muling partner.

The other hybrid fanciers had always aimed for the completely yellow mule. Crossing a yellow canary with a wild finch in the hope of producing a canary-yellow mule is like the alchemists' hope of mixing two base metals to yield gold. Unlike medieval magicians, bird breeders sometimes succeeded at this, but so rarely that these birds were actually worth many times their weight in gold. The goal of producing 'clear' or 'light' mules – brilliant-yellow birds with no dark feathers – is what drives many mule breeders. These are gorgeous birds to be sure, but their beauty lies partly in their scarcity. Rudolf Galloway, one of Davenport's critics, reared hundreds of mules over his lifetime but only one completely clear bird.[16]

Vigorous Hybrids

When Hans Duncker began the quest to breed a red canary in the 1920s a huge amount was known about hybrids.[17] Bird-fanciers across Europe had been generating hybrids for more than 200 years and had written extensively about their obsession. Furthermore, biologists like Darwin and Mendel had also written many pages on the subject. Duncker was therefore well aware that to create his red canary he had to contend with three bits of biology. The first was probably a bonus, but the other two were serious obstacles. First, he knew that any mules he produced would be tough, long-lived birds – a phenomenon known as 'hybrid vigour' and well known to the Egyptians, who were the first to cross horses and donkeys. Second, as was common knowledge among bird breeders, mules are always more difficult to breed than canaries. The third and most significant hurdle was the fact that mules and hybrids are almost always sterile.

The true mule, the hybrid offspring of a horse and a donkey, is unbeatable as a beast of burden, combining the strength of a horse and the endurance and surefootedness of a donkey – unequivocal hybrid vigour. Bird breeders recognised the same phenomenon and in his forty-four-volume *Histoire Naturelle*, written in the mid-1700s, the great but eccentric Comte de Buffon reported that the offspring from the 'cini [serin], siskin and goldfinch with the hen canary are stronger than canaries, sing longer, and their notes are fuller and more sonorous . . .'[18] Even today breeders often comment on the longer lives of their mules compared with parent species.

Whatever bird breeders believe, the truth is that we do not really know whether hybrid finches are 'better' in any way than their parents for no one has bothered to conduct a proper test. Whether a hybrid is more vigorous than its parent species depends very much on the circumstances. When horticulturalists cross different strains of the same species, such as maize, the resulting offspring may indeed show increased vigour, but only because the parent strains were somewhat inbred. In this situation hybridisation restores any variation that was lost when the parental strains were inbred. On the other hand, hybridising very different species may have exactly the opposite effect, breaking up groups of genes that work particularly well together in their proper owners.

Strictly speaking, mule refers only to an animal whose father is a donkey and whose mother is a mare. When the parentage is the other way round – a more difficult feat – the offspring is a 'hinny'. In the past some people claimed they could tell mules from hinnies because their front end resembled the sire and the back end the dam – reinforcing the already prevalent view of male superiority in reproduction.[19] Some fanciers imagined they could see something similar in birds. Buffon related how hybrids assumed the father's appearance in their head, legs and tail, and the mother's in the middle. 'It appears, therefore,' he said, 'that, in the mixture of the two seminal liquors, however intimate we suppose it to be, the organic molecules furnished by the female occupy the centre of that living sphere which increases in all dimensions, and that the mole-cules injected by the male surround and inclose these.'[20]

When bird keepers first produced mules from canaries and native finches, they invariably used a male finch and a female canary because, being domesticated, female canaries were more likely to reproduce in captivity than female finches. Hervieux, however, recommended the opposite: 'For my part I am for the contrary . . . because the male communicates more of his kind to the breed than the female . . . and consequently the mongrels that come from a cock canary-bird will be more beautiful and sing better . . .' This was simple canary sexism. Generations of experience subsequently gave no indication whatso-ever that the appearance of mule offspring is affected by whether their father is a canary or a finch.

Another erroneous belief was something called 'the effect of a previ-ous sire'.[21] The idea that a female's very first sexual partner influenced all her subsequent offspring, regardless of later breeding partners, gained credence with the story of Lord Morton's mare. Lord Morton's aim in life was to save the quagga, a kind of zebra, from extinction, through a captive breeding programme. He failed before he had even started, for he only ever acquired a single quagga – a male. In despera-tion he allowed it to mate with a female horse, which in due course produced a nice hybrid offspring bearing on its back and legs the quag-ga's stripes. When Lord Morton sold the mare and it was later mated to a black stallion, the offspring all bore the tell-tale quagga stripes, indicating – to Lord Morton at least – that the mare's original partner, the quagga, had had an extremely long-lasting effect. Neither Lord Morton nor the Royal Society which published his account in 1820 seemed to have considered that the mare might have produced striped offspring regardless of whom she mated with.

Instead, the idea of previous sires took a firm hold on the imagination of animal breeders and so convinced were they that they even managed to confuse Darwin. He desperately wanted to believe his trustworthy informants who described instances of the effect-of-the-previous-sire, or 'telegony' as it was called – but his biological intuition told him it couldn't be true. Not until the early 1900s, when Mendel's laws of inheritance became common knowledge, was it generally appreciated that *both* parents contribute equally to the production of offspring, allowing biologists and animal breeders alike to stop worrying about telegony. Even so, as late as the 1920s racists hijacked the telegony idea for ideological purposes; in Germany some anti-Semites asserted that 'a single act of intercourse between a Jew and an Aryan woman is sufficient to pollute her for ever. She can never give birth to pure-blooded Aryan children, even if she marries an Aryan.'[22] The final bastion of telegony was among kennel clubs where, curiously impervious to biology, the idea lived on until the 1970s.

Hard to Get

Darwin's notebooks from the 1850s are full of anecdotes, snippets of information and curious facts about hybrids. He believed them to hold the secret to the origin of species.[23] By looking at which species would hybridise and produce fertile offspring and which wouldn't, Darwin thought he could establish boundaries between species. He knew, for example, that the more similar two species were, the more likely they were to hybridise, and that two finch species were more

likely to breed together than a finch and crow. But there were so
many exceptions to this pattern that by the time he put pen to paper
twenty years later, Darwin was unable to infer very much at all about
the origin of species from the study of hybrids. Normally a rapid
writer, he took a full three months to write the difficult chapter on
hybrids in the *Origin*, which he summed up by saying, '. . . crosses
between forms sufficiently distinct to be ranked as species, and their
hybrids, are very generally, but not universally, sterile. The sterility is
of all degrees . . . and . . . does not strictly follow systematic affinity,
but is governed by several curious and complex laws.'

Later, the evolutionary biologist Ernst Mayr confirmed that the
only pattern is a rather messy one. First, it is true that the less related
two species are the less likely it is that they will hybridise. On the
other hand not all closely related species hybridise easily. Whether
hybridisation is possible depends to some extent on whether the two
species naturally occur in the same habitat. If they do then they may
have evolved different courtship displays specifically to avoid hybridisation. Two closely related European warblers, the willow warbler and
the chiffchaff, occur in the same areas of woodland and look very
similar, but they have completely different songs, which enable them
unambiguously to recognise each other and avoid mis-pairing. They
are sexually isolated by song. Such differences evolve because hybrid
offspring are generally less fertile than pure-bred offspring, so natural
selection favours those individuals that breed with their own species.

Some species are in the strange position of being closely related yet
not having encountered each other for millions of years, and this is
precisely the situation of the red siskin and the canary. Both species
evolved from a common ancestor but became separated millions of
years ago when South America drifted away from Europe.[24] Only
when the Spaniards started taking red siskins to the Canary Islands
were the two species united. As a result of their prolonged separation
there were no major behavioural barriers to prevent them from
hybridising.

It is curious that despite the difficulties of producing hybrids many people still believe that most domesticated animals originated from crosses between different species. This widespread conviction stems partly from the fact that generations of selection have rendered domesticated species so different from any wild species that it is often far from obvious what their ancestor was. Early ornithologists thought the canary was a combination of the true canary and the Elba citril finch; others considered canaries a mixture of one or more green European finches – the European siskin, serin or the citril finch. They probably did so for three reasons: they all look superficially rather similar; canaries hybridise easily with these species; their hybrid offspring often look very much like wild canaries. Florence Durham, one of the first scientists to use canaries to study inheritance, was absolutely convinced that the canary had a multi-species origin precisely because her imported wild canaries stubbornly refused to breed with each other, but, perversely, did so willingly with other finches.[25] But canaries, like pigeons, domestic fowl and all domesticated animals, are derived from a single ancestral species. In some cases this has been verified by molecular analysis,[26] but even without the insights of modern technology, the difficulty of producing hybrids and the fact that hybrids are usually sterile, tell us that it is extraordinarily unlikely that our ancestors would have struggled to produce domestic animals by cross-breeding them. Yet this is precisely what Duncker planned to do.

Infertile Hybrids

The more Duncker read, the more he was convinced that the widely held view that all hybrids are sterile was simply wrong. Hervieux, Buffon and Darwin all stated quite clearly that some hybrids – not very many, but some nonetheless – were fertile.[27] He had also heard it said that the red siskin mules produced by Mexican and Cuban fanciers were fertile. Moreover, it was widely known that many plant hybrids were perfectly fertile, allowing nurserymen to produce new 'species' effectively in a single generation. Duncker went back and reread Hervieux: 'The Young Ones that come from these mix'd Birds, often breed others the next Year, contrary to the Opinion of him that has writ the contrary . . .' Duncker also looked again at what Darwin had said[28] using information from the bird-fancier Bernard Brent, that the canary 'has been crossed with nine or ten species of Fringillidae [finch], and some of the hybrids are almost completely fertile'. The important point for Darwin was that these cross-bred birds had not given rise to new species, but for Duncker it was enough that some hybrids were capable of reproducing.

Because hybrids are often sterile, the accepted wisdom is that hybridisation is rare in nature, but this too is a myth. Fully 40 per cent of all plant species are thought to have arisen naturally through hybridisation. For reasons we do not understand, plants interbreed much more readily than animals, a feature horticulturists have enthusiastically exploited. But hybridisation may have played a significant role in the origin of some wild animal species as well. The critically endangered red wolf of North America, it now transpires (on examination of its DNA), is nothing more than a hybrid between a grey wolf and coyote. Birds aren't immune either and a recent survey revealed that 895 species – one-tenth of all known bird species – have been recorded breeding with another species in the wild to produce hybrid offspring.[29] Among captive birds the figure is even greater – 1500 species have been recorded breeding with another species (incredibly, over fifty of these with canaries). A recent study on the Galapagos Islands showed that different species of Darwin's finches sometimes hybridise and that their offspring often reproduce successfully with one or other of the parent species. Just as with plants, hybridisation may – occasionally at least – be a way of introducing new and beneficial genes into animal species, allowing very rapid evolution and the creation of new species, a result Hans Duncker would have found distinctly reassuring.

Some bird keepers disapprove of rearing mules, not because it goes against the will of God or nature, but because they see it as a dead-end exercise. In all other branches of the fancy, breeders select for better birds each generation, creating a sense of continuity and improvement – something completely lacking in mule and hybrid breeding. Undeterred by such views, breeders produce mules and hybrids either for exhibition or for song, but in both cases only males qualify. Females do not have enough colour to make them attractive as show birds and, of course, only the males sing. Most breeders therefore discard their cross-bred females at the earliest opportunity, unless they represent an extremely unusual hybrid. Through a strange quirk of genetics, however, male birds almost always greatly outnumber females among cross-bred offspring.

Buffon was among the first to comment on the unusual sex ratios among hybrids.[30] When he was preparing the canary section of his enormous multi-volume natural history encyclopaedia his friend Père Ignace Bougot the local priest told him of a hen canary paired to a male goldfinch that had laid nineteen eggs giving rise to sixteen male and three female offspring. Buffon quipped that this was 'a greater inequality than ever takes place in a pure breed' – a pattern confirmed by subsequent breeders including the Scottish eye surgeon Rudolf Galloway whose records revealed that of 135 mules he produced between 1904 and 1908 and whose sex he established, no fewer than 110, or 81 per cent, were male.

Funny sex ratios had also been noticed in other hybrids. Among the mixed offspring of donkeys and horses there is always a preponderance of *female* foals – the opposite of what happens with birds. The brilliant evolutionary biologist J. B. S. Haldane was the first person to make sense of these unusual sex ratios. In 1922 he recognised that 'when in the offspring of two different animal races one sex is absent, rare, or sterile, that sex is the heterogametic sex', by which he meant the sex determining the gender of its offspring – an observation that has since become known as Haldane's rule.

In humans and all other mammals, including horses and donkeys, males have two different sex chromosomes, an X and a Y, and when they make sperm, half of them carry an X and half a Y chromosome. Female mammals, on the other hand, have identical sex chromosomes – two Xs – and when they produce eggs, each egg has an X chromosome. When sperm and egg fuse at fertilisation, the resulting embryo always gets an X from its mother and either an X or a Y from its father. If an X sperm penetrates the egg the resulting embryo will be female (XX); if a Y sperm enters the egg, the offspring will be male (XY). So in mammals the male is the 'heterogametic' sex and determines the sex of the offspring. In birds and butterflies things are the other way round and females are the ones with the different sex chromosomes. It was Florence Durham, William Bateson's research assistant, who discovered, during her studies of canaries in 1908 that in birds it was the female who called the shots regarding the sex of the offspring. Bateson was grappling with the same problem at the same time, using chickens, but canaries proved a more tractable system and Florence beat him to it.[31] To his credit, Bateson, generally considered something of a bully and a sexist, allowed her to publish her findings without including him as an author – not something that would readily happen for such a momentous discovery in today's ruthlessly competitive academic environment.

The crux of Haldane's rule is that, among hybrids, if you are the sex with different sex chromosomes you are more likely to be sterile or dead, presumably because when the genomes of the two species fuse at fertilisation, the sex chromosomes prove unwilling partners.

Haldane's rule has huge implications for the origin of species. Once two races or incipient species fail to produce viable offspring, or if their offspring are sterile, they are on their way to becoming separate species. His rule hinges entirely on the incompatibility of the sex chromosomes of different species, but despite eighty years of research, the way sterility or inviability is imposed on hybrid offspring remains a mystery. We do not know whether Duncker was aware of Haldane's 1922 paper, but he certainly knew of Florence Durham's findings, and he knew that among mules, males invariably outnumbered females. What is strange is that in compiling the evidence for his 'rule', Haldane seems to have been completely unaware of the extensive information available from bird keepers – even scientifically respectable ones like Galloway.[32]

Dubious Doppelbastards

The rumours that Duncker was hearing about the existence of red siskin x canary hybrids were true and they started to unfold at exactly the time he began his red canary project. Several fanciers had apparently reared red siskin mules and, just as the Spanish bird keepers had said, some of them were fertile when paired back to canaries. Even more encouraging, Duncker learned that two bird keepers, A. Dams and Bruno Matern in Prussia,[33] had managed to produce 'doppelbastards', that is, double hybrids: offspring from breeding two red siskin mules together.

The success achieved by these amateur bird keepers made Duncker extremely optimistic about producing a red canary and in the late spring of 1926 he began his quest by pairing red siskins to ordinary yellow canaries. This first phase of the project was straightforward. Many of the eggs were fertile, and by the end of that season he and Reich had produced a respectable number of mules. Phase two, conducted the following year, involved pairing male and female mules together – just as Dams and Matern had apparently done. But this time there was no success. Only a few of the birds, mainly the males, showed any sign of breeding, and none of the very few eggs laid were fertile. It must have been a depressing time for Duncker and Reich. And they were puzzled. If others could get mules to breed together why, then, couldn't they? It looked as though, despite all Cremer's funding, they were going to fall at the first hurdle.

Fortunately they persisted.

8

Fugitive Red

Red, the colour possessing the greatest power of attraction, projects the strongest visual energy, as the colour of the picture and the very heart of its being.

JÜRG SPILLER (1962) describing Paul Klee's painting *Blossom in the Night (1933)*

There is no doubt that evolutionary biology has an implicit moral/political message, not least for those who are not trained to guard themselves against these kinds of inferences or do not have an alternative moral framework firmly in place.

U. SEGERSTRÅLE, *Defenders of the Truth* (2000)

Persuaded by Duncker's enthusiasm, Cremer arranged for workmen to start building three new bird houses at Rosenau to be completed in time for the 1927 breeding season. One of the new houses was for Cremer's budgerigars, another for his exotic foreign birds, and the third for Reich's and Duncker's canaries. Cremer's investment extended still further: he employed people to feed, water and clean the birds, and a full-time secretary to maintain the extensive breeding records. Duncker, spared any of the day-to-day responsibility, focused his entire attention on designing the breeding experiments and analysing their results. He was under some pressure to succeed since Cremer's outlay had been considerable and, like all businessmen, he wanted results in return for his investment. But Duncker knew what he was doing. At the very least, by spreading his efforts across three separate projects he bettered his chances of success. And he had a hunch that the budgerigar's kaleidoscopic colour mutations would be more tractable than the canary's variegation genes.

Cremer's wonderful facilities, unfortunately, failed to provide the right ambience to persuade the red siskin mules to breed together. Perhaps Dams and Matern had been lucky with their 'doppelbastards' or perhaps they had been lying – but Duncker and Reich could not persuade male and female red siskin mules to breed together, not in 1927 and not later.[1] The female mules were unwilling participants in Duncker's scheme and seemed to be dead from the neck down. Outwardly quite healthy, they simply showed no sexual stirrings whatsoever. It was as though they had no passionate parts. Duncker dissected a few, putting to good use the anatomical skills he had gained as a student, and to his amazement that is exactly what he found. Instead of a big, bold uterus and a cluster of yellow yolks, these birds were completely devoid of reproductive organs. Little wonder they didn't lay eggs. They were like clocks with no clockwork.

The male mules were different. Several of them were bursting with rude passion fuelled by testosterone churning out of their full-sized gonads. But no matter how motivated the males were, if the females were sterile, there was little point in pursuing the original plan of trying to breed male and female mules together. On the other hand the success of Dams and Matern still rankled. But Duncker had seen enough to realise that even if he persisted and eventually found one or two fertile females, progress would be desperately slow.

Had he been able to reproduce Dams's and Matern's double hybrids, the red canary would have emerged fairly quickly. Duncker's logic came partly from Mendel's pea studies and partly from the practical experience of generations of bird keepers who knew that mules invariably resembled their finch parent rather than their yellow canary parent. In Mendelian terms, the genes for the finch's plumage were dominant over those of the yellow canary. Duncker's research with Reich in the previous two years had shown exactly the same to be true for canaries themselves: genes for the green plumage of the wild-type canary were dominant while the yellow genes of the domesticated bird were recessive.

If he could have persuaded two siskin mules to breed together, some of their offspring – three-quarters, he reckoned – would assume the red cardinal cloak of their siskin grandfather and the rest would be yellow like their canary grandmother. Duncker had also banked on another bit of Mendelian logic: some of the red offspring from mule parents would carry a double dose of red genes and therefore be even redder than their parents.

The female mules stubbornly refused to co-operate. Ever the pragmatist, Duncker modified the original plan and instead of starting with very red mules, he would use the coppery coloured birds he had already produced and back-cross them to canaries. The traditional back-cross involved pairing an individual with its mother, but it wasn't necessary for Duncker to be quite so specific, he merely had to back-cross to a bird that was similar to the mule's mother, a yellow canary. The coppery male mules were half siskin and half canary, and back-crossing these to a hen canary would yield offspring that were three-quarters canary and one-quarter siskin. But he would retain only the reddest birds and cull the rest. In this way he could simultaneously retain the siskin's red genes but expunge nearly everything else that the red siskin had donated to its offspring. The next generation would be seven-eighths canary and one-eighth siskin, but still with red genes. This was the essence of selective breeding and precisely how generations of animal breeders, with no knowledge of Mendelian genetics, had produced wonders like the merino sheep and the cocker spaniel – with one important difference. No one had ever attempted to put the genes of one animal species into another.

Duncker and Reich prepared for the new breeding season by

setting up cage after cage with a coppery male mule and a yellow canary hen. Progress wouldn't be rapid, but the red canary was within their reach. Every day Reich removed and numbered the eggs as they were laid, replacing them with ceramic eggs, or sometimes ivory ones – remnants of an earlier era – placing the real eggs in special trays and keeping them cool until the hens had laid their full clutch of five or six. By returning the eggs to the hen canary only when she had started to incubate, they minimised the chance that the inquisitive male mule might peck them. Then, after four days at 40°C under the hen's brood patch, Reich took the eggs out of the nest again and candled them. By holding the egg up to a strong light – in truth a candle was barely bright enough – he could see whether it was full and fertile with a developing embryo, or clear and unfertilised.[2] It was a frustrating time: only a few eggs were fertile. But Reich and Duncker expected this; they knew it wasn't going to be easy and clear eggs were part and parcel of mule breeding. They threw them away and allowed the birds to breed again, hoping for better luck with the next clutch.

The few fertile eggs were placed under the care of canary foster-parents; old, experienced birds known to be reliable breeders. Fostering is a standard management technique among bird breeders that increases the chances of the chicks' survival. By taking the eggs from the siskin-canary pairs it speeded the production of offspring by encouraging them to lay again. Reich and Duncker waited for the chicks to hatch and for their feathers to emerge so they could see what colour the new hybrids were. It took almost two months for the new feathers to appear and, as every mule breeder knows, the waiting must have been intolerable. Siskin mules were late breeders and the offspring sometimes didn't attain their true colours until November. But when the moult was finally done Duncker was disappointed again. Instead of the Mendelian ratios of red and yellow birds that they expected, these second-generation mules looked much like the first, reddish – bronzy, brassy, coppery, but certainly not crimson. The onus lay firmly on Duncker's shoulders; Reich and Cremer made no claim to understand genetics, the red canary was Duncker's baby and they were looking to him to bring it into the world.

While the red canary was experiencing a difficult gestation, the budgerigar project was in obstetric overdrive, generating results faster than Duncker had ever imagined. Cremer was ecstatic and each evening when Duncker appeared after his teaching day was over, he was updated on the results. The purpose of their collaborative research was to work out what to expect when budgerigars of two different colour types were crossed. There were twelve recognised budgerigar varieties in 1928, many of them with different genetic constitutions; the project therefore required several hundred different test matings, each employing several pairs of birds. As with the canaries, Duncker found that the budgerigar's plumage colour was controlled by a number of different factors and by working out how these were inherited it wasn't long before he could perfectly predict the outcome of any mating. Because green plumage, for example, was dominant, a pure-bred (or homozygous) green bird mated to either a yellow or a blue bird would only ever produce green offspring, and because blue was recessive, two blue parents would only ever produce blue chicks. Other types of pairings produced different coloured chicks in different proportions and large numbers of offspring were needed to check these ratios against the predicted Mendelian values.[3] Such was their enthusiasm for this project that in 1926, soon after they began to collaborate, Cremer and Duncker, with a handful of others, launched the German Budgerigar Society – which continues to this day.[4]

It isn't difficult to imagine Duncker walking home from Cremer's house at dusk along Bremen's leafy streets, thrilled with their results and planning the papers he would write. And just as soon as there were enough chicks from a particular pairing to be sure of the results, Duncker began writing. These were wonderful and exhilarating days, for the budgerigar's colours were, as he suspected, as easy as Mendel's peas. The results were published under Duncker's name, but usually with a note acknowledging that the research had been conducted 'in the aviaries of Generalkonsul C. H. Cremer'. Their findings were quickly translated and transmitted around the world to universal acclaim. For the average budgerigar breeder Duncker had created order out of chaos; breeding budgerigars would never be the same again.

By the end of 1927 Duncker had published over twenty scientific articles on bird genetics. His major discoveries were sent to the *Journal für Ornithologie,* where Erwin Stresemann was the editor. Stresemann was also a canary enthusiast and, rather than dismissing domesticated birds as having no relevance to bird biology, as many other ornithologists did (and still do), he actively encouraged Duncker. Stresemann was smart enough to realise that while ornithologists would dearly love to know about the inheritance of colour in wild birds this was utterly impractical, but by letting Duncker pave the way with canaries and budgerigars, he was building a firm foundation for future researchers.[5]

Duncker also wrote more accessible accounts of his work for bird keepers' magazines like *Gefiederte Welt.* A good populariser, he made every effort to ensure that amateur bird keepers with little or no training in biology could understand his results. Cremer fully supported Duncker in this goal by hosting meetings for scientists and bird breeders at his Rosenau home. In 1927 Duncker decided to establish his own journal with the explicit aim of bridging the gap between scientists and amateurs, and putting the whole business of bird keeping on a more scientific footing. This was no trivial undertaking and *Vögel ferner Länder* (Foreign Birds) became the official organ of the German bird keepers' society – the AZ. Prior to this the society journal had been a thin and poorly produced affair, but in taking it over and renaming it, Duncker transformed it into a top quality publication for his and other researchers' findings. Appearing quarterly, the first volume contained over 200 pages of articles recording its members' attainments. Cremer became the AZ's president, leading the society onwards and upwards with what the membership fondly called his 'golden Rhineland humour'. The society's journal went out to every one of its 423 members. Thanks to Duncker's papers and his visionary editorship, *Vögel ferner Länder* opened new horizons and unimagined opportunities for discovery among Germany's bird keepers. Duncker contributed the first article, 'Inheritance and the Breeder', and no less than five others, including one on his red canary venture.[6]

Eugenic Dreams

On 26 May 1928 Hans Duncker celebrated his forty-seventh birthday. His teaching career was going well and was greatly enhanced by his growing scientific reputation. He was a popular teacher and started a small museum at school based on the skins of exotic birds that died in Cremer's aviaries. Duncker added canaries and budgerigars of different colours from his own experiments in the hope of encouraging some of the boys to take an interest in genetics. At the same time, in the barely overlapping circles of science and amateur bird breeding, Duncker's status was also growing. His prodigious stream of publications was making it clear both to ornithologists and bird keepers alike that Duncker knew more about the genetics of bird colours than almost anyone else.

The period of economic stability that had begun around 1924 was all too brief. By 1929 the political situation in Germany had started to deteriorate once more. Unemployment rose from 1.8 to 2.8 million in a single year; terrorism was widespread and inflation once again raged out of control. Following an aggressive election campaign in September 1930, Hitler's National Socialist party, to everyone's surprise, secured 18 per cent of the vote. The Weimar Republic retained the majority and were still in control, but only just. Thereafter the economic and political situation deteriorated rapidly, aided by the Wall Street crash and by Nazi thugs rampaging through Berlin's streets smashing the windows of Jewish shops. Germany was tearing itself apart economically, socially and biologically.

The German people were restless and receptive to the notion that regeneration could be achieved through scientific knowledge. The idea that society might be improved through an understanding of genetics had been around for at least forty years, introduced under the name of eugenics by Francis Galton.[7] In Britain it remained just an idea, but in the United States eugenic theory had been put into practice during the 1920s and people deemed unfit to reproduce were sterilised – supposedly for the good of society as a whole.[8] Eugenic ideas were also well entrenched in Germany thanks to the popular writings of certain influential biologists, and during the depressing years following World War I, the possibility of improving society – the 'Volk' – seemed especially appealing. Germany felt it could no longer afford the humanitarian luxury of supporting 'degenerates'. Eugenic ideas were rendered even more palatable through crude but seductive economic arguments and the apparent success of similar policies in the United States.

One of the main architects of the eugenics movement on the other side of the Atlantic was the canary geneticist and arch-Mendelian Charles Davenport, whose muddled paper on inheritance of colour and crests Duncker had wrestled with in 1923.[9] Back in 1902 when Davenport had negotiated his position as director of the Station for Experimental Evolution at Cold Spring Harbor, he had told his sponsors that his programme would help in the 'improvement of the human race by better breeding'. Once in place, he started to address the then uniquely American problem of human race mixing. His investigations led him to seek the genetic basis for a wide variety of human traits, including eye and hair colour as well as pauperism, criminalism and the sailor's love of the sea. Some of these studies would be laughable were it not for their appalling social consequences. By the time Galton was knighted in 1909, Davenport had completely embraced his predecessor's idea of improving the human race and was embarked on an ambitious study of 'feeble-mindedness'. He soon became a leading figure in America's eugenic movement and part of the faction that favoured the compulsory steri-lisation of the feeble-minded in order, as Davenport said, to 'dry up the springs that feed the torrent of defective and degenerate proto-plasm'. To halt the degeneration of society, as he saw it, he also advocated the restriction of immigrants into the United States. The notorious Immigration Act of 1924 assessed over 80 per cent of east-ern and southern Europeans seeking to immigrate as 'feeble-minded' and refused them entry. During the 1930s US immigration officials sent back large numbers of Jewish refugees anticipating the Holocaust for the same reason. At the time that Duncker was wrestling with the red canary, Davenport had become president of the International Federation of Eugenics Organisations and was rubbing shoulders with the likes of Eugen Fischer, the German Professor of Anthropology who later provided 'scientific justification' for Hitler's attempted eradication of the Jewish people.

Because the eugenicists believed that differences between races and between individuals were wholly genetic – and hence unchangeable – the process of 'improving' society through selective breeding seemed to them at least both straightforward and highly desirable. Others, however, highlighted the utter fallacy of eugenic beliefs. The former British prime minister, Arthur Balfour, for example, said that if the fit survive, all that means is that those who survive are fit. Consequently the eugenicists' worry that 'the biologically fit are diminishing in number through the diminution of the birth rate' must be wrong by the doctrine of natural selection. . . . If families of the professional class were 'so small that it is impossible for them to keep up their numbers, they are biologically unfit for this very reason'.

In Germany eugenics originated in the late nineteenth century with high-profile scientists like Ernst Haeckel, a committed Darwinian and great populariser of science. Haeckel did more than merely promote evolutionary thinking. He treated Darwin's natural law of selection as a *social* law and declared that different human races represented distinct levels of evolution. 'The lower races such as . . .

the Australian Negroes', he wrote, 'are psychologically nearer to the mammals – apes and dogs – than to the civilized European, we must therefore, assign a totally different view to their lives.' Haeckel's book *Die Welträtsel* (1899) (published in English in 1900 as *The Riddle of the Universe*) sold over 500,000 copies in Germany alone, and led to a much greater misappropriation of science in Germany than in either Britain or America, and it did much to create a uniquely German form of social Darwinism. However, it was their fiercely nationalistic view of society that distinguished German eugenicists from the more liberal, individualist eugenicists of Britain and America.

Haeckel's other popular book *Lebenswunder* (1904) (published in English as *The Wonders of Life* in 1905) anticipated and epitomised the German eugenicist view:

> The history of civilization teaches us that its gradual evolution is bound up with three different processes: (1) association of individuals in a community; (2) division of labour among the social elements, and a consequent differentiation of structure; (3) centralisation or integration of the unified whole, or rigid organisation of the community. The same fundamental laws of sociology hold good for association throughout the entire organic world.

The ideas of Haeckel and his followers, that some races were more worthy than others, that society could be improved and that individuals should work for the common good, were so pervasive that by the 1920s there was, as the political scientist George Stein put it, very little left for National Socialism to invent. 'German academics and scientists did, in fact, contribute to the development and eventual success of national socialism, both directly through their efforts as scientists and indirectly through the popularisation or vulgarisation of their scientific work.'[10]

Duncker was also drawn to the notion of improving the German people through scientific knowledge, especially genetic knowledge. It appealed to his sense of national identity and made him feel as though his scientific skills were serving society. Moreover, his faith in genetics was unshakeable – if his knowledge of heredity and artificial selection had enabled him to improve domesticated birds, why shouldn't the same principles be applied to people? His troubles with the red canary were temporary and he was confident he would soon find a solution.

Without Yellow

The failure of the red siskin mules and yellow canaries to yield the predicted offspring was both disappointing and puzzling. The alleles had not segregated in the straightforward way Duncker expected and he could only conclude that the genes that controlled the canary's yellow plumage somehow interfered with those of the red siskin to create rusty rather than red offspring. Duncker saw it as a battle between yellow and red. If he could only eradicate the yellow, the red would shine through. He thought it all out again and weeks later had an answer. The way forward, he now felt, was to avoid yellow canaries altogether and use white ones, which have no yellow genes, to create a new cohort of red siskin mules. If his assumptions were correct, then in the absence of yellow genes the red would have to predominate.

There were two types of white canary to choose from. Though superficially identical, genetically they couldn't have been more different. One type possessed white genes, which were dominant over those controlling the normal yellow plumage, while in the other the white genes were recessive to the yellow. It was the dominant white genes that Duncker was after.

What were almost certainly dominant white canaries had first appeared in Germany around 1660. Johann Walter painted one and physician Schroeckius, Baron Pernau and Rosinus Lentilius had all written about them.[11] German dealers had carried them on their backs to the Paris bird markets, where Hervieux had seen them. Very few white canaries ever made it to England, but Eleazar Albin, smart operator that he was, had included a picture of one in his book. But rare mutations are always vulnerable to extinction and after a brief period of popularity these white canaries disappeared – the mutation simply died out and it wasn't until nearly 200 years later, in 1908, that white canaries spontaneously re-emerged. Remarkably, they did so simultaneously in two separate stocks about as far away from each other as was geographically possible: in New Zealand and in England. Test breeding revealed, however, that in both cases the white plumage of these birds was recessive to yellow and they were called 'recessive whites'.

Then, in 1918, another white canary cropped up, this time in Germany and out of ordinary green roller canary stock. This proved to be a different mutation, one in which the white plumage was dominant to the usual yellow plumage. Carefully nurtured, this mutation, known as a 'dominant white', was quickly established and proved to be extremely popular. The Germans propagated them so successfully that by 1924 dominant whites were being widely exported across Europe. Meanwhile the recessive white canary had all but become extinct; those in Britain had disappeared completely and only a handful of the New Zealand whites, imported into Europe in 1925, had survived. Duncker managed to see a few of them and realised that the two types actually differed very slightly in their appearance. The recessive birds were pure white, but the dominant white birds had yellow 'shoulders', and we can see from Walter's painting of 1657 and Albin's from 1731 that the original German mutation was also a dominant white bird. It was on this mutation that Duncker now pinned his hopes.

If he used white canary hens and red siskin cocks, Duncker was convinced the canary's dominant white genes would completely suppress any yellow ones and allow the siskin's red to express itself. He also predicted that this pairing would produce mule offspring of which half would be red and half orange (Figure 5). The reason for

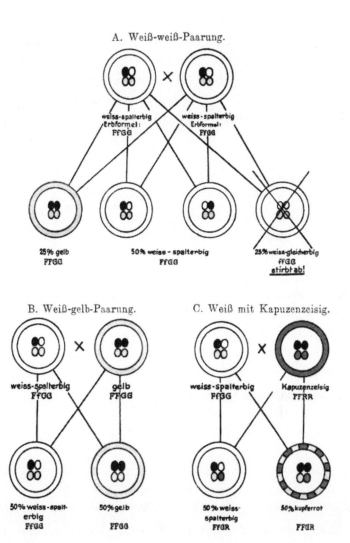

FIGURE 5 *Duncker's genetic schemes for predicting the outcome of particular pairings. Top, the expected results from pairing two dominant white canaries: note how those offspring bearing two white alleles (far right) do not survive; bottom left:pairing dominant white and yellow canaries; bottom right: pairing a dominant white canary and a red siskin.*

166 THE RED CANARY

this was that the dominant white canary cannot exist as a 'pure' mutation, since individuals with two dominant white alleles always die before they become sexually mature (Figure 5, top). All dominant white canaries capable of breeding therefore carry only one allele for white plumage, which is dominant, and a recessive yellow allele, which is not expressed.[12]

More siskins were ordered from South America. When they arrived, Duncker and Cremer set up the pairings with the white canary females and again waited for the chicks to colour up. And for a third time they were thwarted. Certainly, the hybrid offspring were of two colours, but not the red and orange Duncker had predicted. Instead, they were either copper, just like the previous mules, or ash-grey – rather like the female red siskin. Duncker described the season's results in a paper in *Vögel ferner Länder* and included a beautiful colour illustration painted for him by Karl Neunzig,[13] the editor of *Gefiederte Welt*, depicting a male red siskin, a white canary and the two types of very disappointing offspring.

A Captive Audience

Duncker was beginning to lose interest in the red canary. He simply could not understand why his experiments did not work. His logic was impeccable and his approach was scientifically exemplary, but the siskin's red genes were much more difficult to capture than he ever imagined. As we'll see, it wasn't just the genes that were a problem.

Despite the setback with the red canary project, this was still a productive time and 1929 saw the publication of his book *Genetik der Kanarienvögeln (Canary Genetics)*, which became an instant success. It was what canary breeders had always wanted: a scientifically sound step-by-step guide to their breeding efforts. Through the book and his other publications, and as a result of secondary reports in magazines like *Cage and Aviary Birds* in Britain and the *American Cage*

Birds, news of Duncker's research reached out around the world. His discoveries were an inspiration to canary breeders everywhere, even inside the American penal system, where they encouraged Robert Stroud – the 'birdman of Alcatraz', incarcerated alone for fifty-four years – to create his own coloured canaries.[14] Duncker's book allowed canary breeders for the first time to predict with complete confidence the kind of offspring they would get from pairing particular combinations of birds. Now that breeding canaries to order was relatively straightforward, fanciers wanted a new challenge and the red canary project captured their imaginations. Fanciers around the world launched themselves into it with unbridled passion.

9

Not by Genes Alone

The bombs fell indiscriminately on Nazis and anti-Nazis, on women
and children and works of art, on dogs and . . . canaries.

CHRISTABEL BIELEBERG, *The Past is Myself* (1968)

As the news of Hans Duncker's pioneering efforts spread across
Europe and the world, more and more bird keepers were caught
up in the rush to breed a red canary.

Of the thousands who took up the challenge during the 1930s only
a handful have retained a place in history. One of these is an
Englishman, Anthony Gill. Well educated and well read, Gill not
only infused new life into the project, he also started the Red Canary
Movement, in 1947 renamed the Canary Colour Breeders' Association,
which continues to this day.[1] Gill was one of a handful of enthusiasts
who saw the operation through almost from its beginning to its final
flowering. A dapper little man, a librarian by profession, Gill wrote
with elegant simplicity out of deep knowledge. His breeding experiments were well publicised in the cage bird magazines and later
summarised in a unique book in the 1950s. By then there were dozens
of books on canaries but, because plagiarism had continued unabated
since Olina's days, most were poorly researched, poorly written and,
in anticipation of small print runs, poorly produced. Gill's book,
New-coloured Canaries, published in 1955 when he was eighty-one, is
among the best books on cage birds ever written.

In the early 1930s, as Gill was beginning to enjoy the warm flush of excitement in anticipation of producing a red canary, Charles Bennett, a research biologist on the other side of the world in San Francisco, was experiencing similar emotions. Born in the same year as Duncker, Bennett earned his place in the red canary story by being the first to recognise that redness lay not just in the genes but also in a canary's diet.[2]

There was a key player in Prussia, too, who like his homeland has retained only a shadowy place in history. Bruno Matern of Rastenburg in central East Prussia, whom we met earlier, was almost completely eclipsed by Duncker, yet he played a pivotal role in creating the red canary. Matern's involvement began in 1912 with a single bird. A certain Herr Engels of Tilsit, having bred some red siskin x canary mules, gave one to a Herr Dams, a railway worker from Königsberg, who succeeded in back-crossing it to a female canary. Several young birds were produced from this pairing but they all died and in 1914 a disheartened Dams gave the mule, which was obviously fertile, to Matern and wished him good luck. An experienced bird keeper, Matern succeeded, against all the odds, in producing some orange canaries from this single cock bird. When Duncker got wind of this in 1925 he was impressed by Matern's success and later generously acknowledged him as the true red canary pioneer. Dams was also astonished by the skill with which Matern propagated his solitary male mule and in 1926 he travelled across Prussia by train – probably on a free ticket since he was an employee – to Rastenburg to see the results for himself. Face to face with Matern's most recent cohort of coloured canaries Dams was ecstatic, declaring them the 'most beautiful birds in the world'. He wrote an account of his visit for *Gefiederte Welt* to celebrate Matern's genuine genetic success for, as he was keen to point out, these orange birds had not been colour-fed. Touchingly, he also recounted how, as he prepared to leave, Matern gave him one of his precious orange canaries as a gift; a gesture that moved Dams to tears.[3]

Natural Allies

As the passion ebbed from Duncker's red canary quest, his reputation as a bird geneticist continued to grow. In February 1927 he and Cremer were awarded special gold medals by the fledgling British Budgerigar Society 'for their great and unselfish work on behalf of the budgerigar' and they were invited to the Crystal Palace bird show in London in 1928 to collect their honours. Unfortunately they couldn't go, because Duncker was in ill health and had to undergo a major kidney operation. Instead, the medals were sent and Cremer wrote to thank the Budgerigar Society on behalf of them both.[4] By the following year Duncker had bounced back, greatly encouraged by the enthusiastic reception of his new book *Kurzgefasste Vererbungslehre für Kleinvögel-Züchter (Inheritance for Bird Breeders)*, which he dedicated to his loyal friends Reich and Cremer. Invitations to speak at meetings around the world now started to arrive thick and fast, and Duncker was honoured by being made vice-president of the British Western Counties Budgerigar and Foreign Bird Society. In February 1930 he and Cremer went together to the National Cage Bird Show at the Crystal Palace in London as official guests of the British Budgerigar Society. After admiring and judging the British budgerigars, Duncker met Anthony Gill, then vice-president of the British White Canary Club, and took the opportunity to scrutinise the white canaries that were on exhibition, discovering to his amazement that every single one was of the dominant white form. Later that year, again accompanied by Cremer, Duncker gave talks in Vienna, Tübingen and at the International Ornithological Congress in Amsterdam. But when he was invited to Australia and then to the USA for the 6th International Congress of Genetics at Ithaca in 1930 he couldn't afford to go – presumably because Cremer, who probably paid his expenses, had to draw the line somewhere.

Yet attempts within Germany officially to recognise and reward Duncker for his extraordinary success fell on stony ground. In

November 1930 his director at school, Herr Jentsch, wrote to the Bremen education committee:

> After reading in the newspaper that the Senate has appointed Dr Sachs a professor, I feel motivated to ask you if Dr Duncker could be honoured in the same way. A detailed explanation for my suggestion will hardly be necessary, as Dr Duncker's name is widely and internationally known in science through his research in the area of genetics . . . if perhaps you would be willing to support my suggestion . . . I would be extraordinarily grateful.

Less than a week later Jentsch was informed by the education committee that making Duncker a professor would not be appropriate because it would 'contradict paragraph 4 of article 109 of the constitution of the Reich'. Bureaucrats!

Duncker's supporters tried another tack. The next month geneticist Alfred Kühn, Professor of Zoology at the University of Göttingen and successor to Ernst Ehlers, Duncker's old supervisor, wrote asking the Bremen Senate to consider creating a position for Duncker as an independent researcher. They replied:[5]

> I would like to express my thanks to you and your fellow signatories of the letter from 12th December 1930 for the interests you have in the scientific works of Dr Duncker . . . Of course it is also my urgent desire to enable Dr Duncker to continue with his valuable scientific work. Unfortunately there could hardly be a more difficult moment to create an independent research position for Dr Duncker or even to substantially reduce his teaching duties. I certainly do not have to explain the general financial pressure on the Reich, federal states and communities. We are finding it most difficult to provide even the most essential funding for education . . . We will do our best not to put too much pressure on Dr Duncker. In particular we will try to employ Dr Duncker at school only in his special disciplines and not

in his secondary subjects. A reduction of total teaching hours however is not possible at present.

Kühn didn't give up. He went to Bremen's mayor to discuss the possibility of finding a partial replacement for Duncker to allow him more time for research. But a note appended to the above letter, dated 13 February 1931, said that although negotiations were continuing, there was little hope of success.

So Duncker carried on with his teaching and continued trying to understand why the dominant white canary had failed so spectacularly to produce any red mules. He decided that the dominant white genes must in some way block the expression of the red ones – as the yellow genes had done in the earlier experiments. Still convinced that white canaries held the key, Duncker was now certain that the birds that would generate the right results were those with 'recessive' white genes – hence his interest in the Crystal Palace birds. Years later his hunch about white canaries turned out to be correct, although not in the way he imagined. Now, however, instead of persuading Reich and Cremer that they should do the test matings themselves, Duncker merely published a short article late in 1929 proposing recessive white canaries as the solution.[6] His enthusiasm was fading, for although he was one of only three breeders in the world with any of the precious recessive white canaries, he simply couldn't motivate himself to test his ideas.

The only other person in Europe with recessive whites was Anthony Gill.[7] A bird keeper since boyhood, Gill was seven years older than Duncker and in his mid-fifties when he committed himself to produce a red canary. The year before Duncker published his paper predicting on recessive whites, Gill had purchased some of the descendants of Duncker's red siskin x canary hybrids, making him the first person in Britain to own reddish canaries. But it is not clear – to me at least – exactly when he decided to pursue the red canary. Still, reading Duncker's paper the next year must have confirmed

Gill's conviction and when the two men met at the Crystal Palace bird show in February 1930 and it became clear that Duncker wasn't going to do the experiments himself, Gill saw the green light. Years of experience and a ready grasp of the laws of inheritance, combined with an eye for detail and meticulous record keeping, made Gill confident he could succeed.

It took two years for Gill to build up his stock of recessive white canaries and to acquire some red siskins from a contact in Spain. By the spring of 1933 Gill was ready to start. He knew exactly what he had to do: punctilious as always, Duncker had been utterly explicit about what he expected. Because these canaries carried no yellow genes whatsoever, the siskin's red plumage would dominate and all the offspring would be red.[8]

But they weren't. Six months of optimistic nurturing and patient expectation left Gill completely crestfallen. The mules produced by his recessive white canaries were either copper-coloured cocks or ash-grey hens, depressingly similar to those from Duncker's yellow canary mothers. Gill consoled himself that this was progress of a sort, for it now seemed abundantly clear that Duncker's hypothesis of a conflict between red and yellow was incorrect. If the recessive white canaries carried no yellow in their genetic constitution, there was no way in which red and yellow genes from different parents could squabble over precedence. Late in 1933, when the new mules had developed as much colour as they would get, Gill reluctantly decided that Duncker must be wrong. His emotions were mixed; on the one hand he was full of admiration for what Duncker had started, but he also recognised that the ball was now very obviously in his own court.

Gill abandoned the white canaries for a while and sent for some of Bruno Matern's orange wonders. He put these to good use, crossing them with male copper hybrids and producing birds of a deeper orange than ever before. They weren't red, but their more intense colour encouraged Gill and others to continue the quest. In Prussia, Bruno Matern was pursuing much the same kind of experiment and, like Gill, had noticed that instead of always producing deep orange canaries, the coppery mule and orange canary parents occasionally and completely unexpectedly threw some very odd-looking white offspring. These ghostly individuals always turned out to be hens and most breeders gave them away, thinking that they were of little value in their crimson crusade. But Matern was an experimenter and had crossed one of the white hens back to a coppery hybrid, and discovered to his absolute astonishment that the resulting offspring were a deeper orange than any birds he had produced so far. More exciting still, after two or three generations the females of these orange canaries proved to be fertile – accelerating the production and selection process enormously. Matern's deep-orange birds eventually gave rise to a strain that he called 'Rastenburg Reds', even though they were still basically orange.[9]

Further confirmation that there was no contest between red and yellow genes came from Charles Bennett's research. Bennett had been one of the first people in the United States to get his hands on imported orange birds in the 1930s and breeding coloured canaries was what Bennett did in his backyard during evenings and weekends. The rest of the time he was a physiologist at the University of California at Berkeley. It was largely due to Bennett's efforts that the red canary fantasy was soon raging across America as well as Europe. Indeed, as A. K. Gill noted, the passion for the red canary there was even greater than in Europe: 'Nowhere has its effect upon practical canary breeding so extensively manifested as in the United States.' Enthusiasm was fuelled by Bennett's articles in cage bird magazines and later by his 1945 orange canary book. Other books by American fanciers followed in quick succession, reflecting the increasing fascination for coloured canaries.[10] Lucius Armitage published *Science in Color Breeding* in 1947 and a year later came Herman Osman's *The Orange Canary* based on pieces he had written for the magazines *Canary Journal* and *Canary World*. In addition, associations specialising in coloured canaries sprang up across North America. It was as though canary breeders had been waiting for something exciting to happen and, when it did in the form of an orange canary, their enthusiasm for their hobby was revitalised.

The popularity of coloured canary projects in North America and Europe meant that from the 1940s huge numbers of red siskins were taken from their native Venezuela to fuel the bird breeders' fantasies. Most fanciers focused exclusively on the red siskin as the source of red genes, but a few enterprising individuals experimented with other red birds, including the California linnet – also known at the time as the Mexican rosefinch and, more recently, the house finch which despite producing fertile mules were never red enough to make a lasting contribution. One of the most unexpected alternative red birds was the African red fire finch which a Mr Howard Lewis, also of San Francisco, had successfully hybridised with a canary back in the 1920s. That Lewis had used his hybrid offspring to produce a strain of orange canaries was truly astounding, for the fire finch and canary are only very distant relatives and the chances of their hybridising at all, let alone producing fertile offspring, were extremely low.[11] Bennett went to see Lewis's birds and was impressed by their vigour but, as he was quick to point out, they were distinctly orange and therefore another blind alley as far as the red canary was concerned. Lewis reared more than 1000 of his vivid orange hybrids but their fame was short-lived. By the end of the Second World War they were extinct.

Bennett was responsible for the crucial turning point in the red canary story. The obsession with *breeding* a red bird had blinded Duncker and thousands of others to the possibility that, on their own, genes for redness might not be enough. What they overlooked, ironically, was that some of their imported red siskins eventually changed colour, from vermilion to copper. The evidence that redness required more than mere DNA was staring Duncker and his followers in the face, but Bennett was the only one to see it.

The trigger for Bennett was the revelation that *people* sometimes changed colour – from pink to yellow – after eating certain foods.[12] A condition known as carotenemia was recognised initially in four women who for reasons best known to themselves had eaten four pounds of raw carrots a week for seven months and turned bright orange. The same symptoms could also result from eating too much spinach, or too many eggs, as one man found after having decided 'without medical advice' to live on nothing but eggs: three weeks and 567 eggs later he too had turned an unhealthy shade of orange.

The substance that gives carrots their orange colour is carotene – one of a family of compounds known as carotenoids, first identified in 1886. The colour in eggs comes from a different carotenoid, lutein, and experiments conducted in the early 1900s revealed that chickens fed a lutein-free diet produced eggs with white yolks. Giving them carrots had no effect on their eggs, but feeding them lutein restored the yolks' colour.

The realisation that the secret of colour lay in the diet sent Bennett scouring the university libraries for information. And he found it – with almost every page he turned Bennett discovered more and more evidence that colour in nature was influenced by diet.

One of his main sources of inspiration was an extraordinary book published in 1893 by Charles Keeler, *The Evolution of Colors in North American Land Birds*. Keeler was a pioneer and his thesis was that the colours of birds had evolved through the Darwinian processes of natural and sexual selection. At a time when Darwin's ideas were out of fashion, Keeler was well and truly ostracised for his intellectual audacity. Reviewing the book for the American bird journal *The Auk*, Joel A. Allen – the editor and a founding member of the American Ornithologists' Union – described Keeler as 'seriously handi-capped . . . by his lack of experience and familiarity with exotic birds'. Allen couldn't stomach Keeler's 'hap-hazard conjectures' – he wanted facts not ideas, especially from an uneducated upstart who 'has still his spurs to win in the field of zoological investigation'. In a marvel-lous and carefully measured response Keeler pointed out that without new ideas science would stand still and that because Allen – like so many ornithologists – was firmly stuck in Lamarck's muddy legacy of inheritance of acquired characteristics, it was hardly surprising that he didn't countenance Darwinian ideas. Following Allen's mean-spirited review, Keeler's book[13] more or less disappeared – but fortunately, not before Bennett got to see it.

Keeler made Bennett aware that rather than being distinct and antagonistic, yellow and red in a bird's plumage were a continuum: 'When red occurs in a group of birds, yellow will be found in the same group, and that the yellow represents a more primitive stage of development than red.' For Bennett the evidence was overwhelming. Hadn't the red siskin got a sister species, the saffron siskin found elsewhere in South America, identical in every respect except that the red was replaced by yellow plumage? Weren't there birds like the scar-let tanager, in which the male was red and the female yellow? And didn't some birds, like the California linnet, invariably fade from red to yellow in captivity? The two colours had to be related. The more Bennett thought about it the more it seemed like a natural progres-sion, just as fruit starts out green and turns yellow, then orange, before ripening to red.

Bennett's professional research interests gave him access to the rather specialised scientific literature and he was delighted to come across a 1934 article by two German researchers, Hans Brockman and Otto Völker, which showed unequivocally that diet could affect the colour of canaries.[14] Their article appeared in a journal that most bird keepers would never have heard of and it described how yellow canar-ies deprived of carotenoids turned white after their moult: their new feathers were completely devoid of colour. These researchers also found that their birds' yellow colour could be restored, but only with lutein – the very same substance that put the yellow back in egg yolks.

Bennett and Gill started to write to each other and their ideas began to converge. Colour, they decided, must be determined in part by a bird's diet. To verify this Bennett embarked on some experi-ments of his own in 1946. His method involved no more than plucking a few feathers from his birds' breasts and allowing them to regrow while the birds enjoyed either an experimental or a normal diet. Using this primitively pragmatic technique, Bennett could perform experiments throughout the year without having to wait for his canaries to undergo their annual moult. Brockman and Völker had used only yellow birds in their experiments, and Bennett was particularly interested to see what effects carrots would have on his *orange* canaries. To his delight, the carrot-fed birds regrew feathers more richly coloured than their original orange plumage. He was breaking new ground, Then, to check that it was carotene and nothing else in the carrots that turned the birds a deeper shade of orange, Bennett repeated the experiment but this time supplemented the birds' diet with carotene oil rather than carrots. Again the new feathers emerged a deep orange. Intriguingly, when he fed *yellow* canaries carotene oil, their new plumage remained virtually unchanged. Finally, Bennett performed the reverse experiment, withholding carotene from orange canaries and found the new feathers to be white.

A picture was emerging. Carotenoids had a more dramatic reddening effect on orange canaries than on yellow ones. Bennett captured the essence of his findings in his book *A Study on the Nature of the Orange Canary* by saying, 'A yellow bird will not be made red by any amount of carotene, and the degree of redness which an orange bird can acquire from the ingestion of carotene is sharply limited by its heredity.'[15]

Eureka! Genes and diet were working in unison to produce an orange canary.

The discovery of something like the secret of the canary's colour after a long hard struggle is what the Germans call the 'egg of Columbus' and Bennett had cracked it.[16] He had identified what is now referred to as a gene-environment interaction – one of the most fundamental phenomena in biology because it recognises that the nature *versus* nurture debate is erroneous and meaningless. Despite the overwhelming importance of this interaction many people persist in talking about traits like human intelligence, weight or height as though they were determined exclusively by genes *or* by the environment, or as if some fixed percentage of these traits was determined by genes and the rest by the environment. For red canaries, colour is determined both by the possession of particular genes *and* by an environmental effect – the presence of carotenoids in the diet. Red genes without carotenoids cannot produce a red bird and conversely, as Bennett showed, no amount of carotenoids will turn a canary orange or red if it lacks the right genes.

Bennett's excitement was understandable, but his appreciation of the gene-environment interaction in his red canaries was hardly novel. Mendel's predecessor at the Brno monastery in Moravia (now the Czech Republic), Cyrill Napp, was well aware of the general idea. Moravia was the centre of scientific sheep breeding and wool was an important source of income for the Brno monastery. Abbot Napp, an extraordinary scientist and breeder of both plants and animals, was one of a group of sheep breeders who questioned whether the Spanish merino sheep would retain the quality of its wonderful wool in different environments. Their concern was based on the knowledge that particular breeds of sheep did best in particular pastures – highlands, lowlands, wet and dry environments.[17] Yet despite predating Bennett by a century, the idea of gene-environment interaction remained an elusive and poorly grasped concept, and one easily eclipsed by the more polemical nature-nurture debate.

What about the red-pepper effect with which Bemrose had caused such a brilliant fuss seventy years previously? The vital carotenoid in peppers, Brockman and Völker revealed, was 'capsanthin' and it worked differently from carotene: even in the complete absence of red genes it could turn a yellow canary orange. This was why, from the moment Bemrose had gone public in 1873, the breeders of Norwich canaries all started to feed their birds peppers. Ever hopeful of producing even deeper orange birds, breeders continued to experiment, and throughout the late nineteenth and early twentieth centuries they concocted elaborate recipes from sweet peppers, Natal peppers and paprika together with such ingredients as brown sugar, biscuits and olive oil. They didn't make much additional progress. Advocates of colour feeding considered it a 'harmless, innocent and honest' way of bringing 'hidden treasures to the surface'. Others, less keen, worried about the long-term effects of colour feeding, especially hot peppers, on their birds' health. But it was pointless to abstain. Using peppers to enhance their birds' plumage had become such a central part of the exhibiting culture for certain types of birds that those who didn't colour-feed didn't have a chance of a prize.[18] Bemrose's controversial discovery had very clearly shown that substances in a bird's diet could affect the final colour of its plumage. But what he had not recognised – and this is hardly surprising for his era – was the link with heredity. Charles Bennett's major contribution some seventy years later in the 1940s was identifying the link between genes for redness and diet.

With more than a tinge of embarrassment but also with mounting excitement, Bennett and Gill realised that red canaries had existed undetected for years simply because no one – including Duncker – had appreciated that redness, even in naturally red birds like the red siskin, depended on dietary carotenoids. Bennett and Gill called their birds 'red factor canaries' – and 'factors of heredity' was the expression Mendel had used for what we now call genes. Red factor canaries were those carrying the genes that, with the right diet, could turn them a deep orange or red. So obsessed were Duncker, Reich and Cremer with *breeding* a master race of red canaries that they never contemplated colour-feeding their coppery mules. But the genetically pure race of red birds that Duncker dreamed about simply could not exist. It was a biological impossibility.

Difficult Diets

The pivotal role of environmental carotenoids in colouring birds was discovered in captivity, but unbeknown to Gill and Bennett the very same experiment had been conducted among wild birds – albeit unwittingly and in a remote corner of the world: in Hawaii. The first Europeans to reach the west coast of North America were reminded of home by a small red singing bird they named the California linnet. The male bird sports a grubby red head, breast and rump, while the female is entirely dull brown. Though not nearly as red as its southern hemisphere cousin, the red siskin, the California linnet – or house finch as it is now known – has provided considerable insight to the business of being red.

Everywhere human colonists have gone in the world they have felt compelled to take their familiar native species with them. The British excelled at this and, in what became a massive acclimation industry, transported all manner of birds and mammals by their hundreds to the Antipodes in the mid-1800s – with a devastating effect on the local fauna. Americans did the same when they settled in Hawaii in 1870. Among other possessions they took with them house finches from the Pacific coast. Only a few years later the ecologist Joseph Grinnell was shocked by the change in appearance of these birds on Hawaii. Instead of being bright red they were now a dingy orange colour. Their metamorphosis was so remarkable that Grinnell reckoned they had mutated into a completely different species, and promptly rechristened them Hawaiian linnets and gave them a new scientific name – *Carpodacus mutans*.

It subsequently became apparent that house finches also varied considerably across the North American continent, with birds only a few kilometres apart sometimes appearing radically different. Much later, in the 1990s, when researchers fed orange birds from Hawaii a carotenoid-rich diet during the moult, they were amazed to see them turn a rich red colour. The Hawaiian birds had not mutated. They had retained the potential to turn red, but in the wild they didn't, because naturally occurring carotenoids were scarce in Hawaii.

There is an extraordinary parallel to this story, involving canaries in the same part of the world. Soon after the Pacific Cable Company set up a base in 1909 on Midway Atoll – the desolate northern outpost of the Hawaiian chain – the governor's wife released eleven yellow canaries there. Later, visitors to the island reported that these free-flying canaries, which had increased enormously in numbers, were no longer yellow but had become white. They had done so for the very same reason the house finches on Hawaii turned yellow – a lack of naturally occurring carotenoids.[19]

Birds cannot manufacture carotenoids on their own. If the compounds aren't part of their natural diet, birds cannot develop yellow or red plumage. Other essential substances, like testosterone or estrogen, are synthesised in the body, but not carotenoids. It may be precisely for this reason that red plumage is such a big deal in the bird world – at least to female birds. What better clue to a male's vitality than his ability to garner rare carotenoids from the environment? Given a choice (in aviary experiments) between a dull male and a red one, female house finches go for the red guy every time. It is the same in nature: when biologist Geoff Hill, now at Auburn University in the USA, compared male house finches with and without a partner, the paired males were much redder than the bachelors. More convincing still, when he dyed wild males to turn them either more or less red, those made redder were much more likely to find a partner.[20]

Redness presumably reflects a male's foraging skills, an attribute of crucial importance to females seeking a partner with whom to raise a family – equivalent to women judging a man's earning power by the car he drives. But house finch redness says much more than any auto-mobile, since it also reflects a male's health. Hill found that birds infested with feather mites were less brightly coloured than unin-fected birds and those unfortunate individuals suffering from avian pox – a viral infection currently sweeping like the Black Death across North America – were also less red than pox-free birds. By choosing the reddest male she can find, a female house finch is getting a healthy provider – the best of all possible worlds.

Red plumage in male house finches has evolved because females preferentially mate with red males and it works as a barometer of male quality precisely because, unlike men's cars, it is an honest signal. A man can hire a car or go badly into debt for one, but a male house finch has no short cuts to colour. In most areas carotenoids are hard to come by and sick or weak males simply cannot fake red plumage. Darwin would have been delighted by Hill's findings. Charles Bennett, incidentally, anticipated Hill's study by forty years when he noticed that one of his red siskins became sick just before its moult and 'when the moult was over, it came out the palest adult male Siskin we had ever seen, with some feathers showing only a weak orange colour'.

Among the orange house finches of Hawaii we can guess that females choose the brightest partners they can find, since these represent the best of what's available. One can also imagine that if we were to import a few red males to Hawaii, the females would go crazy fighting for their favours, but so far no one has done that experiment. Geoff Hill did something similar: he took Hawaiian females to Canada and gave them a choice of different coloured males in aviaries, where they reacted precisely as predicted. They went for red males – males that were redder than any they had ever seen. Although the environment had altered the colour of Hawaiian males, it hadn't changed the females' preference for deep red.

What about regions where carotenoids are relatively abundant – equivalent to a situation where every young man owns a Porsche? Again, we don't know the answer, but one of the several species of Asian rose finches provides a clue. Pallas's rose finch, which inhabits the windswept steppes of eastern Siberia, has gone one better than most other species: in addition to its gorgeous crushed-strawberry

plumage, it also bears a glitzy silver necklace. My guess is that carotenoids are not too difficult to find in Siberia and the females of Pallas's
rose finch are forced to use an additional criterion – the possession of
throat feathers that glitter like shards of polished steel – to distinguish
good from indifferent males.

I also suspect that female red siskins assess potential partners in
much the same way as house finch females – the redder the male's
plumage the better, but because siskin males vary so little in colour,
females add in extra tests, such as song quality. We might never know
if this is true, because the red siskin is all but extinct in the wild.

Blood and Mud

The political backdrop against which Duncker's acclaimed publications and successful bird projects took place was a violent one.
Destitute and disillusioned, Germany thrashed from left to right in a
desperate effort to reinvent itself. As the historian Michael Burleigh
pointed out,[21] for many 'Nazi ideology offered redemption from a
national ontological crisis; to which it was attracted like a predatory
shark to blood . . . a climate of hopelessness and despair tempted
people to support the National Socialist movement. . . . All people
had to do was to make the quantum leap of faith; unified national
self-belief was the solution to every mundane problem.' Nazism's
appeal spanned much of German society offering a 'cool and highly
reasoned approach to reality based on the greatest of scientific knowledge', which gave racism a respectable gloss and which in turn
seduced the 'ascendant intellectual force of the day.' Nazism was politics as 'a biological mission, but conceived in a religious way.'
Duncker, and dozens like him, welcomed it.

I did not discover Duncker's Nazi links until I was more than
halfway through writing this book and it came as a terrible shock. A
German friend had been helping me research Duncker's past and we

had arranged to meet outside the Natural History Museum in London. It was there that he broke the news. Sitting on a bench in front of the museum in the warm spring sunshine, he handed me the article he had discovered. It was written in 1990 and published in the journal of Bremen's Natural History Society – the very society Duncker had served for so long. Hubert Walter, Professor of Human Biology at the University of Bremen, had exposed Duncker's Nazi allegiance and pronounced him a disgrace to biology for so enthusiastically promoting their eugenic ideology. Walter documented the evidence for Duncker's leading role in Bremen's society for racial hygiene, including the lecture programmes he arranged for local National Socialist party members, his own influential lectures on eugenics, and his reluctance to recant or 'apologise' after the end of the war – as some others had done.[22]

This was a bitter blow. I had started this book with the intention of making Hans Duncker a hero – the lost pioneer of bird genetics, forgotten by ornithology and ignored by the rest of science because he worked entirely on domesticated birds, the antithesis of what real scientists studied. Now at least it was clear why he had been overlooked. Although disappointed, I was not entirely surprised because I knew that a large proportion of German academics had joined the National Socialist party. Some did so to preserve their careers without any ideological commitment; but Duncker was apparently not one of these. His enthusiasm for Nazi ideology was known to some of those who worked with him after the war, but the extent of his involvement had not previously been generally known. Walter's paper made public Duncker's Nazi links and Walter singled him out for condemnation because his enthusiastic support for eugenic ideas brought both German ornithology and the Bremen museum into disrepute.

It took two years to track down a photo of Duncker and when I finally found one, taken in 1913, he was in his prime and looked friendly and respectable. After months of researching Duncker's background, discovering his Nazi involvement was like a bombshell. I vacillated. I looked for excuses – I found plenty, for eugenics was by no means restricted to Germany – other European countries and the USA thought it was a good idea too. But no one, of course, took eugenics to such morally repugnant extremes as the Nazis and Duncker had been part of this atrocious process. Later, I found another photograph of him, taken around 1939, when he was in his late fifties; hard-faced, tight-lipped and sporting a Hitleresque moustache.

Biological Abuse

When the Nazis seized power in 1933 Duncker found himself in sympathy with much of their ideology, which he and many other German biologists had long shared. The alacrity with which he embraced the new regime was typical of his approach to life. He was an enthusiast and his rising passion for the new politics nudged the failing red canary project into second place. We do not know whether Reich and Cremer were as keen as Duncker on the new government.

Duncker had been elected vice-president of the Natural History Society in 1931. Two years later, together with his close friend Hans Meyer, the society's president, he had started a sub-unit for racial hygiene. This occurred partly because the Natural History Society had long considered the study of man an important part of its remit, but also because radial hygiene was a topic of increasing interest. Soon after coming to power, the National Socialists announced their plan to introduce a law requiring the genetically unfit to be sterilised, and Duncker and Meyer arranged a series of public lectures at the museum to discuss the plan. The five lectures, under the title 'The Prevention of Unworthy Life', took place over several weeks during March and April 1933 and were given by a biologist, a sociologist, a psychiatrist, a theologian and a professor of law.[23] The proposed sterilisation law wasn't due to come into force until early 1934, so the lectures allowed people time to discuss and adjust to the idea. Contrary to what one might imagine, these weren't the strident, tub-thumping performances we now associate with Nazism. Instead, they seem almost reasonable in the quality of the discussion and the range of opinions they present – although only the theologian categorically disapproved of the sterilisation plans. It is the utter plausibility of these eugenic arguments that now make them seem so sinister. These ideas were not unique to Germany. At one time or another Winston Churchill, Julian Huxley and the biometrician Karl Pearson had all expounded eugenic views regarding 'unworthy life'.[24]

Duncker continued to edit and produce his journal *Vögel ferner Länder* and he published the odd article on the red canary, but if you run your eye along the entire set of *Vögel ferner Länder* on the library shelf you can almost see Duncker's enthusiasm draining away. From 1932 the number of pages declined with each succeeding volume. Duncker was contributing fewer and fewer articles, and additional editors were needed to help produce the 1934 and 1935 volumes. After 1933 his main interest seemed to be supporting the new government. Part of the reason for this, surely, is that one of the things Hitler did on coming to power was to elevate biology as a school subject, putting it on a par with German language, history, religion, geography, gymnastics and military sports.[25] The change inspired the Dunckers to move house, and in 1933 the family took up residence on the fash-ionable and architecturally elegant Mathildenstrasse, which was closer to the school and to the museum. Although not in the most impressive building on the street, their first-floor apartment at No. 78 was a definite step up in the world.

Just as Duncker was losing interest in the red canary project, others in Germany were beginning to pursue it with increasing vigour. Colour was a national obsession, not least because for the past thirty years German industry had been at the forefront of dye and pigment research. The spin-offs from this effort splashed across German society, creating a synergism between industry, science and art. Biologists interested in the microscopic structure of animals and plants were quick to exploit coloured dyes to stain their specimens, including chromosomes, so named because they are literally 'coloured bodies'. The palettes of Paul Klee and other Bauhaus painters were revitalised and ignited by the vibrant array of new colours available. Pigment research also spawned colour theorists like Wilhelm Ostwald, a consultant to the paint industry and the winner of the 1909 Nobel prize for chemistry. Ostwald devised an ingenious, practical system for comparing colours based on a three-dimensional scale originally dreamed up by the American artist Albert Munsell, who defined colours in terms of hue, value and chroma (that is, their type, such as red or yellow, their lightness or darkness, and their intensity respectively). The system consisted of a series of coloured chips against which anyone, anywhere could compare virtually anything. Instead of having to rely on subjective names like 'crimson lake' or 'alizarin', '*blasskressrot*' or '*hellkress*', for the first time people could make standardised, quantitative measurements of colour.[26] Duncker had been using Ostwald's scheme for years to score his birds. It is a sign of how popular coloured canaries were becoming, and how seriously breeders viewed them, that in 1938 Duncker persuaded the German canary magazine *Kanaria* to issue a special set of colour chips, so that breeders could score their orange and red birds on a common scale.[27]

In that same year the eighty-year-old Cremer was crushed by a car as he crossed the road in front of his home on the Schwachhauser Heerstrasse. He lingered for several days, before finally succumbing to pneumonia. His death marked the end of a unique and spectacularly productive collaboration with Duncker as well as an enduring friendship.[28]

War

On 3 September 1939, in response to the German invasion of Poland, Britain and France declared war on Germany. Much to his surprise Duncker, now fifty-eight, was called up, although he didn't see any action because the decision was reversed almost immediately. Bremen weathered the early stages of the war nearly unchanged but once the Allies' bombing raids began, younger children, including those from Duncker's school, were sent away to rural areas for safety – the so-called *Kinder-Landverschickung* – where they were spared the horror of bombardment. In the meantime the authorities appropriated the Lettow-Vorbeck School and the remaining boys were taught in a building on Dechanatstrasse not far from the cathedral in the old part of the city. British bombers targeted the town quickly. Not only were Bremen and nearby Bremerhaven vital North Sea ports of immense strategic importance, they were also centres of German armament production, with U-boat, Focke-Wolf aircraft and military vehicle factories. In 1935, when Hitler visited Bremen accompanied by the usual display of orchestrated military might – hundreds of troops marching in the plaza outside the railway station beneath flapping red Nazi banners – the predominantly left-wing factory workers gave him a distinctly frosty welcome. Bremen was far too liberal to be a hotbed of Nazi nationalism and Hitler never returned there.

The air raids on Bremen began in earnest in 1942 and the staff at the Übersee-Museum moved the precious specimens to the cellars. As the raids started, one Bremen resident was prompted to comment, 'If ever a city was prepared for disaster it was Bremen.' At the sound of the air raid sirens people rushed to the concrete bunkers, like the one near the railway tunnel at Schwachhauser Heerstrasse. Those who lived in well-built houses on streets such as Mathildenstrasse were able to take shelter in their own cellars as Hans and Elsa Duncker did. I wonder if they reminded each other, as they listened to the sound of falling bombs, of Göring's pledge that no enemy planes would cross

Germany's border.[29] But there were casualties despite the town's preparations, and the inmates of the nearby concentration camp – some of them interned since 1933 – were made to clear up the bodies and rubble after the raids. In Berlin, also a focus of the Allies' bombs, Erwin Stresemann moved his museum's 40,000 bird skins together with Röting's paintings and other treasures into an underground bank vault for safety.[30]

It was during Bremen's ninety-first raid, on 4 June 1942, that Reich's shop in Fedelhörenstrasse was destroyed. Like many other canary breeders Reich had more or less given up his birds during the war because he couldn't get enough food for either himself or his canaries. He was already living in temporary accommodation and the loss of his business was the last straw. He left Bremen to live in Coburg, close to where he had been born. The war extinguished Reich's canary passion and his name never reappeared in the AZ membership list. Worse still, he and Duncker probably never saw each other again.[31]

As they began to push home their attack, the Allies kept Bremen firmly in their sights as a point of invasion. The relentless bombing destroyed much of the city, killing over 3000 people. By April 1945 it was all over. With no electricity, no food and no stomach for any more bombs, many of Bremen's 280,000 inhabitants breathed a sigh of relief as the first British troops appeared. Duncker was devastated. The proud German nation had failed again and the enemy paraded through the ruins of his beloved city.

Prior to the Allied invasion Churchill, Roosevelt and Stalin had decided that as soon as Germany agreed to unconditional surrender, they would 'remove all Nazi . . . influences from public office and economic life of the German people'. The denazification process began with the tracking down of those responsible for war crimes, but it quickly extended to thousands of others suspected of Nazi involvement. Well known in Bremen as a public speaker with Nazi links, Duncker was immediately suspended from his teaching

position, as were many other teachers. Before the war the Nazis had
recruited teachers into the party precisely because of the influence
they had over young people. Now the Allies were determined to
destroy any remaining influence Nazi teachers might have. Duncker
was also on the Allies' list for questioning for his contacts with other
suspected Nazis. The interviews took place in his former school on
the Hermann-Böse Strasse.

Here is what Duncker told the Allies. In 1933, soon after the Nazis
seized power, he was offered the position of director of the Kaiser-
Wilhelm Institute for Genetics in Berlin-Dahlem. This was the
position of Carl Correns, one of those who rediscovered Mendel's
work in 1900. It was a prestigious post and Duncker was deeply flat-
tered but there were strings attached, since to accept the position he
had to join the Nazi party. He declined on two counts. First, he didn't
want to be seen as an opportunist and second, he felt the position
should go to a professional scientist rather than to a teacher who did
research only as a hobby. Even though he had refused to join the
party, in 1934 the government gave Duncker the job of vetting all new
biology books to ensure that they were consistent with Nazi ideology.
Duncker was given this position on the strength of the biology textbooks he had written, but the post of 'inspector of biology books'
didn't last long, Duncker said, because it involved too much work. At
about the same time someone at school denounced him to the Nazis
for making insulting remarks about Hitler. Inevitably there was an
investigation, but because the accusation proved to be greatly exaggerated there were no disciplinary measures. Nonetheless, the incident
had serious repercussions: Duncker was sacked as a night-school
teacher, his promotion prospects disappeared and he was actively
discouraged from giving public lectures.[32]

In 1935 Duncker's superiors at school tried once more to persuade
him to join the party, but again he refused, even though it was made
clear that this would cost him his promotion and make life difficult
at work. It didn't, Duncker said, because he was on good terms with

most of his colleagues. All the same, outside school things became increasingly unpleasant. Clearly frustrated by Duncker's reluctance to join them, the Nazis struck where they knew it would hurt and appointed additional editors to Duncker's journal *Vögel ferner Länder*, diluting his editorial influence. As if this weren't enough, the Nazis also amalgamated the AZ, the bird society Duncker had resurrected and built up, with another bird club, the DWV (Deutscher Waldvogel-Verein – the German Forest Bird Club) and gave Duncker's position as director to a party member. The dramatic decline in Duncker's contribution to *Vögel ferner Länder*, which I had initially attributed to a voluntary change in interests, now made sense in the light of the Nazis' pathetic persecution.

Perhaps inevitably, Duncker's application in 1936 for promotion to senior master at the school was rejected by the local Nazi party leader, who wrote in a letter dated 11 August, 'We recommend the temporary postponement of the intended promotion of master "Studienrat" Dr Hans Julius Duncker, Bremen for one year. His attitude towards National Socialism does not convince us that he is internally totally dedicated to our movement. Therefore we think that a blocking period is still essential.'

It took a further three years for Duncker to secure his promotion and, as he pointed out to his interrogators, it happened only because the Nazis no longer insisted on approving promotions. The following year he was once again under pressure to join the party. Again he resisted, saying that he had never been politically active. Then, in 1940, when they presented him with a pre-completed form that simply required his signature, he capitulated and signed.

Duncker admitted to his inquisitors that he had always believed in racial hygiene, but had never shared the extreme attitude the Nazis subsequently developed. In 1933 he had felt very positive about the new racial hygiene laws that focused on the sterilisation of mental defectives and the encouragement of large, healthy families, but he opposed anti-Semitism and said that he had never personally discriminated against Jews. As the war went on, Duncker said, he became increasingly disillusioned by the Nazi party and felt they were hypocritical in advocating family values while breaking up families by sending men to war. He added that he was sorry he hadn't been involved in the development of the racial hygiene laws – if he had, things might have been better. On the other hand, seeing the way things turned out, Duncker acknowledged it was unlikely that he would have had much influence.

Asked about his involvement with the racial hygiene society in Bremen, Duncker described how the society, of which he was director in 1933, might have gone the same way as his bird society and journal under the Nazis, had he not made it a sub-unit within the Natural History Society and persuaded Richard von Hoff to serve as its chairman. The powerful and influential von Hoff – who had earlier offered Duncker the position in Berlin – proved an effective protector but an idle leader and Duncker continued to do all the work for the society. He recounted how, as a member of the Office of Race Politics, he had organised publicity for eugenic ideas and he described how out of place he felt as the only non-party member at an eight-day course on public speaking organised by the Nazis. Completion of the course allowed him to give public lectures on eugenics, which he did throughout the war. It was precisely this public promotion of eugenic ideas that Hubert Walter later exposed and condemned.

The Allies classified Duncker as a *Mitläufer*, a Nazi hanger-on. It is difficult to know whether the slant Duncker put on his eugenic activities in the interview was entirely honest. Under the circumstances it seems obvious he would have tried to distance himself from Nazi views, yet most of what Duncker said is corroborated by other sources. Moreover, his account makes it clear that had he acquiesced and joined the Nazi party earlier, he would have saved himself a great deal of trouble. After the war Duncker resolutely refused to discuss his beliefs with anyone, which explains why it was possible for Walter, writing in 1990, to say that Duncker had never recanted. Walter described Duncker as an intelligent scientist who was able to explain complex scientific issues in an accessible way. What Walter found hard to understand was how such a critical thinker could have been so uncritical in his support for racial hygiene. 'Duncker', he wrote, 'was one of the very many Germans who readily accepted and propagated the aims of the National Socialists' racial politics and hence contributed to the fact that these aims became a cruel and deadly reality for many human beings.'[33]

On reading Walter's account of Duncker's activities I feared the worst. A Nazi was a Nazi. But Duncker's response to the Allies' interview showed that there were different degrees of Nazi involvement. By his own admission Duncker was a firm eugenicist, although not necessarily an extreme one. He had long been interested in human biology and he told his interrogators that when his budgerigar research came to an end due to lack of funds in 1935, his interests shifted to genealogy. During the 1940s, at the request of the Bremen committee (Witt Heit) he began tracing the social, biological and economic development of Polish immigrant families who came to Bremen in 1886. Duncker spoke on this research at a meeting he organised in 1942, but the study was never finished and the material was lost during the war. It is not clear what motivated this study of Polish immigrants, but given the Nazis' obsession with detecting Jewish blood in previous generations, at the very least they would have approved of it. Duncker said he opposed discrimination against Jews and never personally discriminated against them. Nonetheless, in a lecture given in 1941 he wrote, 'Some nations, like the Jews, may destroy a state but can never build up a state. The only time Jews built up a state was in the time of King David, but the Jews always tried to influence the build-up of their host nation's state – we know that only too well. Fortunately for us that is a thing of the past.'[34] This is clearly anti-Semitic, but whether Duncker believed this to be true or whether he included it to please his audience is something we cannot know.

It was because Duncker was such a charismatic and effective public speaker that Walter condemned his popularising of eugenic ideas. The titles of Duncker's wartime lectures embraced a range of nationalistic topics: 'The Body of the Nation and the *Volk* is built on families', 'The Importance of Grandchildren' and 'Purity of Race Secures the life of a Nation'. After discovering the transcripts of these lectures in the Bremen library, Michael Birkmann, an ex-pupil of Duncker's, felt that Duncker must have been a Nazi rather than a mere 'hanger-on'. Hubert Walter evidently had a similar opinion. However, neither of them knew what Duncker had said when interviewed by the Allies, nor, apparently, of his resistance to joining the Nazi party, or of the Nazis' harassment of him, all of which suggests that Duncker was not the extreme uncompromising Nazi Walter's account implies.

It is impossible to establish how far Duncker's views coincided with Nazi ideology. The problem offers us a glimpse of the difficulty the Allies faced in ranking Nazi suspects after the war. My feeling is that Duncker was a fervent nationalist, a social Darwinist and firm believer in eugenics as a way of improving the German nation. He was probably more than a *Mitläufer*, but with very clear views and clear boundaries with regard to eugenics, neither of which he felt obliged to discuss after the war.

Fluttering Decoys

Born in Germany – albeit as an ugly duckling – the red canary virtually disappeared from that country after Duncker abandoned it. A few diehards, like Julius Henniger, one of Duncker's most avid followers, attempted to revive it, but there was little enthusiasm for a red canary in post-war Germany. The true German canary was the roller, a singing bird, not a coloured one.

But in Britain and North America, thanks largely to Anthony

Gill and Charles Bennett, the red canary – or at least orange canaries carrying red genes – became increasingly common. Its wild counterpart and gene-donor, the red siskin, was rapidly disappearing from its native South America, however. Since the 1890s huge numbers of red siskins, dead and alive, had been exported to the European capitals – as millinery accoutrements and cage birds, respectively. During the late 1800s there was a resurgence of birds' bits as fashion accessories and red siskin skins provided a jaunty and eye-catching patch of fiery colour. But it was the red siskin's willingness to hybridise with canaries that generated a boom in live imports in the first decade of the twentieth century. Initially most birds were imported into Germany for the likes of Matern, Engels, Cremer and Duncker, but as the red canary craze spread, they were sent elsewhere in Europe and to North America.

Relentless trapping decimated the red siskin populations.[35] In the 1940s, under pressure from two local ornithologists, the Venezuelan government made it illegal to trap and sell them – the first time a bird had been given legal protection. But instead of restricting trade, this legislation merely alerted the world markets to the birds' scarcity and in response trappers redoubled their efforts – such was the passion to create a red canary.

Venezuela's new law forced the red siskin trade underground, but enforcement was hopelessly inept and smuggling ludicrously easy. Moreover, because the Netherlands refused to sign the Convention on International Trade in Endangered Species (CITES) dealers were able to launder thousands of red siskins and other birds through the Dutch-owned Caribbean island of Curaçao. The red siskin was in a dreadful downward spiral and the rarer it became the more red canary breeders were prepared to pay for one. The species became so rare in the wild that in July 1975 it was classed as 'endangered' and placed on Appendix 1 of CITES, but even this didn't staunch the flow of these beautiful birds. Sadly, a great deal of wheel reinvention went on, with hundreds of bird breeders during the 1940s, 1950s, 1960s, 1970s and

1980s buying red siskins to replicate Duncker's and Gill's experiments in the hope that a miraculous mutation would deliver them a redder than red champion. In the 1980s, long after the red canary had been put to bed, male siskins were still fetching $1000 each and providing an enormous incentive to trappers and dealers alike. Only in 1987, when the Netherlands finally signed up to CITES, did the trade start to subside. By then it was almost too late, for by this time there were no more than a few hundred birds left in the wild, making the red siskin one of the rarest birds in the world. Conservation has been almost completely ineffectual in securing the birds' future. Today, there are probably a thousand times more red canaries in the world than wild red siskins.

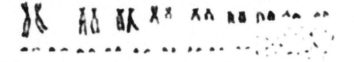

10

Honest Red?

That a red canary will eventually be produced can hardly be doubted. It may be obtained gradually as a result of continued selective breeding. It may appear suddenly as the result of a chance recombination in one individual of the responsible siskin genes.

A. K. GILL, *New-coloured Canaries* (1955)

In February 1948 Duncker was allowed officially to retire and receive a pension – albeit a modest one. He was sixty-seven. Still highly motivated, he appointed himself honorary curator at the Übersee-Museum with the aim of restoring the neglected bird collections. Among the museum's treasures were thousands and thousands of dried bird skins accumulated by a previous curator, the ornithologist Gustav Hartlaub, who for sixty years until his death in 1900 had visited exotic locations around the world simply to shoot, collect and stuff birds. Despite the efforts of Duncker and others to protect it, this valuable collection had been damaged during the war and, like much of Germany, was now in a rather sorry state. Duncker threw himself, with typical energy and enthusiasm, into the task of salvaging, cataloguing and preserving these precious specimens.

For several years after the war had finished, conditions in Germany remained fairly desperate. Bremen was a city of ruins and basic commodities continued to be scarce. Duncker was so strapped for cash he couldn't even afford to pay his subscription to *Gefiederte Welt* and wrote to the editor excusing himself. In November 1948 he received a very welcome food package from an American ornithologist, Jean Linsdale, who was director of the Hastings Natural History Reservation in California. Deeply touched, Duncker wrote to thank him and described how, on discovering that the parcel contained coffee, he and Elsa had quickly made some – savouring their first cup for several years! He also said how much they had appreciated the chocolate. Twenty years previously, Duncker and Linsdale had exchanged journals: *Vögel ferner Länder* in return for the *Condor*. But Duncker hadn't had a copy of the *Condor* since 1939 and as for his own journal, 'it stopped during the Nazi time', he said, using the derogatory term '*Nazizeit*' – something that might have been hard to explain had I not seen the account of his Allies interview.[1]

By 1950 things were starting to improve. In the museum Duncker had almost completed the huge task of restoring and relabelling the collection of 16,000 bird skins. As I discovered when I visited the museum's basement storerooms Duncker had relabelled every single specimen in his fine, scientist's handwriting. In the museum's logbooks Duncker made a master list of all the extant bird specimens and calculated the percentages of each family that had survived the war. In among the cupboards of specimens I found a dusty pair of red siskins, one of Bruno Matern's red siskin mules, and a large collection of canaries from Duncker's early variegation work with Reich. There were even examples of the infamous recessive and dominant white canaries that had let Duncker down so badly in his red canary quest.

Duncker now retrieved the bird specimens that he had previously donated to the school museum. This may have been because after being dismissed he didn't feel that the school deserved them, or perhaps because there was no one there who wanted them. To ensure their safety he took them to the Übersee-Museum. Many of these species, including the red siskin, have since become extremely rare in the wild, making them important additions to the museum's collections.

Cressrote Number 17

Duncker's success in restoring the bird collection to something close to its former glory was widely appreciated, both by Helmuth Wagner, the museum's new director, and ornithologists elsewhere, including Erwin Stresemann, still in Berlin but now on the wrong side of the Iron Curtain. Duncker's efforts were rewarded in 1951 when the Übersee-Museum celebrated his seventieth birthday by making him a member of honour.[2] This might have been a double celebration, for by this date the red canary was as good as established – thanks to Gill, Bennett and dozens of others – but as he made clear in a lecture he gave around this time, Duncker did not see it like this at all. His speech was, naturally, on inheritance, and he used the example of *his* red canaries to show how certain traits were inherited: 'When you cross a canary with a red siskin you get offspring which are *"cressrote"* in colour, and their offspring in turn were also red.' He continued, 'It was never possible to achieve the dark red of the red siskin – the best we could achieve was *"cressrote* Number 17" on the Ostwald colour chart. . . . The English breeders', he said, 'produced birds of identical colour, but' – and this was the key for Duncker – 'only by giving them cayenne pepper. Their red colour was not inherited.' Obsessed with genes to the last. As far as he was concerned, colour-feeding was cheating – his aim had been an honest one: to *breed* a race of red canaries.[3]

During the early 1950s some of the other researchers at the museum noted Duncker's extraordinary dependence on the director's approval and how, if Wagner failed to call in on him during his daily rounds – which started at 7.30 a.m. – Duncker became upset. Duncker's dependency may have been due to the need to feel wanted, but it may also have been because many of the younger researchers at the museum were desperately trying to distance themselves from their country's earlier abuse of biology. One of these, Gerd von Wahlert – later professor – had an office on the same corridor as Duncker and in 1954 was asked by Wagner to write about the educational role of the natural history museum. In doing so he took the opportunity to condemn the Third Reich's horrifying perversion of biology. Duncker's Nazi links were well known, but no one ever discussed this with him. As von Wahlert later told me, others in the museum who had been Nazis both before and during the war had recanted and were 'pardoned', but because Duncker never did he was never formally forgiven.[4] This didn't prevent Hans and Elsa Duncker inviting von Wahlert and his wife for 'a spoon of soup' at their old-fashioned apartment in Mathildenstrasse, which in the typical Bremer's understated way turned out to be a full-blown, formal three-course lunch. When von Wahlert gently teased Duncker about his smoking habit, Duncker replied proudly that there had never been a day since his confirmation that he had been without a cigar. As von Wahlert cast his mind back almost fifty years to recall his lunch with Hans and Elsa Duncker, he suddenly remembered that there were budgerigars in their living room.

Duncker must eventually have paid his subscription to *Gefiederte Welt*, for in 1950 the editor, Joachim Steinbacher, wrote requesting his help to settle a red canary argument. The issue was one of terminology: what should the increasing number of red and other colour varieties be called? Scientists avoid confusion in wild birds and other animals by using internationally standard scientific names, but no such system exists for the varieties of domestic animals produced by artificial selection. The potential for confusion is therefore considerable, as is the potential for individuals to gain status if their particular terminology is adopted. It was precisely this type of conflict that Steinbacher was seeking to resolve. The battle was between Julius Henniger, Duncker's most avid follower, and Karl Hotter, who was jostling with him for the position of top red canary man in eastern Germany. Exactly what one called a colour canary was not a trivial issue, for by the 1950s the number of colour mutations was rapidly increasing. Anthony Gill listed more than ten varieties with names that are dreadfully confusing to the uninitiated, including 'orange greens', 'red orange greens' and 'dilute red orange greens'. The Germans had their own names, mostly stemming from Ostwald's colour scheme adopted during the *Nazizeit* when foreign terms, like 'orange', were illegal. The argument Duncker was asked to arbitrate was whether the German colour canary breeders should adopt Henniger's or Hotter's system. If a particular breed of canary should be called *Blasskressrot* (Henniger's pale-nasturtium-red) or *Hellkress* (Hotter's bright nasturtium) now seems a minor issue, but for Henniger and Hotter this was a battle for enduring status, for whoever's system was adopted would gain a certain immortality.

One step ahead of Hotter, Henniger had already written to Duncker, making a case for why *his* terminology should have precedence and, for good measure, denigrating Hotter for having been a Nazi. We don't know what Duncker told Steinbacher, but it seems he must have sided with Hotter, for when Henniger privately published his book on colour canaries ten years later he complained bitterly that other fanciers had failed to adopt his system of names.[5]

Even in retirement Duncker carried on giving public lectures. Now they were on the biology of birds, chromosomes and inheritance rather than eugenics. Judging from reports in the local newspapers these performances were a tremendous success[6] – Duncker was still an engaging and effective speaker – and the topics he chose were of great general interest.

There are few extant remnants of Duncker's personal life, but some lie in the museum's archive. 'Archive' sounds rather grand, and smacks of order and organisation, but in reality it comprises a broken box file of unsorted papers and the view it provides of Duncker is no more than the sort of glimpse we get through the window of a house as we pass by on a bus. Perhaps the single most important artefact is the yellowing typescript of a lecture on human psychology and its inheritance that Duncker gave in 1950 in which he drew on his own family as an example. The idea Duncker introduced in this lecture was that people fall into one of two types; outgoing individuals who work for 'life' – whom Duncker calls *Lebenwerkmenschen* – and 'day workers', those who are more inward-looking and concerned with day-to-day matters of domestic life – *Tageswerkmenschen*. His implicit assumption was that these two personality types were genetically determined like the colour of ordinary canaries; one dominant, the other recessive. His father, a judge, a local politician and founder of a conservative newspaper company in which the workers had a share in the profits, was a life person. His mother, on the other hand, was a day person, keeping house and helping her husband. Their offspring, as expected, were a mixture of the two types. His younger brother inherited his mother's personality and was a 'day person' and became a pastor in a small village. His elder brother was a mixture of the two types, while Duncker himself was like his father – a dominant life person. To his dying day Duncker's view of the world was one of uncompromising genetic determinism.

Throughout his seventies Duncker visited the museum almost every day. Anthony Gill's book *New-coloured Canaries* was published in 1955 and it seems likely that Gill sent Duncker a copy. Duncker would have been pleased with the prominent role accorded him in the creation of Gill's red factor canaries, but disappointed that Gill was satisfied with birds that relied on colour-feeding. Moreover, because Gill had much closer links with American canary breeders, and because he recognised the essential input carotenoids had had in creating the red canary, Gill dedicated his book to Bennett rather than Duncker. Two years later, at the age of eighty-three, Gill was dead. In the same year Helmuth Wagner wrote a short account of Duncker's life for the Bremen Natural History Society journal, applauding his early scientific work on heredity and his recent restoration work at the museum, but also alluding to his eugenic publications – the cue, I assume, for Walter's later exposé.

In 1960, the year the Beatles first performed in Hamburg, Duncker's wife Elsa died. Even after she had gone, Duncker – now nearly eighty – continued to visit the museum. Then, in September 1961, almost a year to the day after his wife's death, he became ill. Unable to care for himself, he was taken to live with one of his daughters at Saarbrücken on the French border and it was there, on 22 December 1961, that Duncker died.

Internal Strife

In England the red canary society started by Anthony Gill thirty years previously soared from strength to strength and his book continued to serve as a beacon for those still striving for an even redder canary. For many fanciers the fire of enthusiasm still burnt brightly, if somewhat less vigorously, for the birds seemed to be as red as they were ever going to get despite the combined efforts of thousands of fanciers. They were reluctant to admit it, but the red factor canaries they had were still not the crimson colour they all dreamed of and the more sceptical fanciers considered even the very best birds to be no more than deep orange.

In the years just before he died, Gill had been frustrated by the fact that so few members had remained true to the original crimson objective. Many of them, he grumbled, satisfied with orange or pastel birds, had branched off on their own down less colourful but more hopeful avenues.[7] Nonetheless, they all continued to experiment with dietary supplements in the hope of enhancing their birds' colour, for competition on the show bench remained sharp. Contestants had to abide by the club's rules, however, which ever since 1947 had decreed that they could feed their birds anything in its natural state that could be grown in the garden. This ruling tested the ingenuity and resolve of bird keepers, many of whom were not gardeners and some were periodically panicked by the discovery of a fellow fancier cultivating something exotic on his allotment. High in carotene, Russian comfrey had a brief period of popularity during the 1960s with some breeders swearing by its colour-enhancing effect. But in reality nothing had much influence. Certainly red pepper, which could turn a yellow canary orange, added little extra colour to a genetically orange bird. The only trick that seemed to have an effect was to avoid giving birds anything containing yellow carotenoids during their moult, for it was firmly believed that this reduced the intensity of any red colour. Accordingly, fanciers avoided feeding their moulting birds on canary seed and dandelion leaves (both rich in yellow carotenoids) and instead maintained them on a deadly dull diet of niger seed and oatmeal – free of any yellow carotenoids whatsoever.[8]

After years of deep-orange stagnation the red canary took a sudden and brilliant leap forward in 1964. In that year, to everyone's amazement, the national club secretary, Jack Swift, and his two best mates, exhibited some startlingly red birds whose plumage bordered on the elusive crimson they all fantasised about.[9] There were gasps of amazement at these haemoglobin-red birds and intense speculation about what they must have been fed to produce such remarkable plumage. Flushed with success, Swift and his team toured the length and breadth of the country collecting prizes and carrying away silver cups wherever they went. It was like rerunning the Edward Bemrose video – except now the future was even brighter. Frustrated losers huddled in corners or behind closed doors, cogitating and gossiping about Swift's meteoric rise to fame. They pestered him relentlessly for his secret, but as long as he continued to fill the glass-fronted cabinets in his living room with trophies Swift, like Edward Bemrose before him, kept quiet. Swift's extraordinary success was not without its glitches, however, and at a few shows experienced judges like George Lynch became suspicious and ruled Swift's birds out of order, refusing to consider them on the grounds that they must have been fed on something very obviously not grown in the garden. Disqualified and deprived of the prizes he felt he deserved, Swift was bitter in his recriminations – he was the club secretary for God's sake! But many admired Lynch's firm stand, and at shows where Swift's brilliant birds won and deprived others of what they felt was *their* due the losers were vitriolic in their condemnation. The furore wasn't confined to the schoolrooms and village halls where these canary competitions took place, nor to the smoke-filled pubs where exhibitors went afterwards to celebrate or commiserate. It soon became public, and the pages of *Cage and Aviary Birds* crackled with controversy as Swift and his supporters came under increasing criticism.

At the end of the 1964 show season, Jack Swift, who was not a young man and somewhat overweight, was taken ill and rushed to hospital. He remained there for weeks getting weaker and weaker, his condition undiagnosed. Unlikely to recover and unable to continue his duties as secretary, the club reluctantly decided that they had no choice but to replace him. When a new secretary was found, Swift's wife dutifully handed over all the club's files and paperwork, unaware that in doing so she was putting the final nail into her husband's coffin. A week later as Bill Newland, the new secretary, sorted through the files he was thunderstruck to see what his predecessor had been up to. There in front of him in black and white was Jack's secret – a receipt for a new carotene product with the trade name of Carophyll.

Carophyll had been created by the German pharmaceutical giant Roche,[10] and Swift had got to hear about it in 1963 at the corn merchant's where he was employed. Many of his customers were poultry men and one of them told him about a new food additive for chickens that coloured their egg yolks deep orange. Swift persuaded the man to sell him a small amount – it wasn't cheap – and he took it home to feed to his orange birds. The scruffy envelope that had been surreptitiously passed to him at work made him feel like a criminal, but when he looked inside and saw the smattering of coarse maroon dust, he wondered which of them was the crook. He was told that the chicken men simply mixed the dust with the birds' mash, so this is what Swift did, although he had no idea how much to use. Before presenting his orange birds with their new diet, Swift caught them and, using his forefinger and thumb, pulled a few breast feathers from each one. Swift felt more than ever like a criminal, but like willing accomplices, the birds didn't even flinch – the feathers almost fell out of their own accord. It was like planting magic beans, and Swift waited impatiently for the patch of replacement plumage to grow. It took several weeks, but as the new feathers pushed out of their protective sheaths, Swift could hardly believe his eyes – they were as close to a red siskin's in colour as he had ever seen. Carophyll, like Viagra, had revitalised the canary breeders' flaccid libido and engorged their extended phenotypes with an intensity of colour never seen before. The effect was so startling that Swift didn't dare show his wife. Instead, he confided in his two closest friends and between them they agreed to keep their seductive miracle a secret. Perched precariously on the threshold of success, they pledged to keep the vivid birds they subsequently produced under wraps until the next show season. Carophyll would make them famous.

Carophyll is the trade name of a carotenoid with a cumbersome if not wholly unutterable name of 'canthaxanthin' (pronounced can-tha-zan-thin) and was developed specifically to put the colour back into the eggs of battery chickens – it still is. The same substance was also sold in tablet form in the United States as a tanning agent, but was withdrawn after someone died as a result – albeit in a blaze of colour.[11] Orange feathers on a bird look beautiful but bright-orange skin – or to be accurate, the fat beneath the skin – on a human looks positively alien. Nonetheless, dissolved in water (it looks like blood) or mixed in with their food during the birds' moult, Carophyll transformed birds from orange to red – providing, of course, they carried the crucial red genes.

Newland lost no time in revealing the secret of Swift's seedy success to the club's senior members. They were incandescent. Not least because it explained why earlier in the year, in what was now clearly an abuse of his privileged position, Swift had proposed (unsuccessfully, as it happened) that fanciers should be allowed to feed *anything* to their birds. Wallace Dean – a top canary judge – was infuriated by Swift's behaviour because, having bought some of his red birds for a hefty sum, he then watched them moult into ordinary orange canaries.

That Swift doctored his birds was hardly novel. Newland knew dozens of fanciers who had tried it. The motivation to win, whether on the canary show bench, at the canine equivalent, Crufts, or at the Olympics, is so strong that the temptation to cheat is sometimes irresistible. Just as athletes take the red blood cell booster Epogen (EPO) to enhance their performance, so canary breeders pumped their birds full of blood-red pigments to increase their chances of one winning. In every branch of the bird fancy cheating was rife. It was so bad in the poultry world that there was even a book describing the devious tricks some exhibitors got up to.[12] Faking poultry was easier than faking canaries or mules, but it still happened and bird keepers told me how (as young men new to the fancy) they had been duped by linnets smeared with lipstick or by white canaries washed in detergent to increase the reflectance of their feathers.

Surprisingly, not all canary fanciers were outraged by Swift's use of Carophyll. Some saw it as a much-needed advance because it was now possible to produce truly red birds. Others felt it was a betrayal of everything they had worked for over the past forty years. After much debate the club officials finally agreed that since it was impossible to prove whether a bird had been fed on Carophyll they had no choice but to allow its use. In their turn, the members now had no option but to use it if they wanted to remain competitive. This was red raw sexual selection, comparable to a situation in the wild where a new super-attractive male mutant arises, driving females to mate with it to the exclusion of all others.

The saga had gone full circle for canthaxanthin was a dye and feeding it to canaries was nothing less than dying them from within – no different from feeding red pepper to yellow canaries, except that Carophyll in combination with the right genes was infinitely more powerful. Some breeders justified its use by saying it simply brought to the surface a bird's natural attributes, but in reality Carophyll is a potent and vigorous dye, which both enhances and exaggerates a bird's natural colours. But canthaxanthin was hardly something canaries would encounter naturally, for in nature it occurs only in the exoskeletons of saltwater shrimps. Canthaxanthin allows flamingos to retain a rosy hue in the wild because shrimps are what they eat, but it is certainly not what keeps red siskins, house finches or any of the other red finches red in the wild. At least, not directly. What seems to happen among small red birds is that they acquire natural carotenoids – possibly from berries or particular seeds, no one really knows for sure – which are transformed in their bodies into canthaxanthin, which is then pumped into the growing feathers. Feeding birds neat canthaxanthin, as Swift did, and as all red canary enthusiasts have done ever since, bypasses all the normal metabolic pathways and allows huge amounts of red pigment to be dumped directly in the birds' plumage. One way to think about this is to imagine the bird as a Christmas tree. If it has red genes in its make-up, the tree has light sockets in which coloured bulbs can be placed. If you put red bulbs in the sockets – equivalent to feeding the bird Carophyll – the tree lights up red. Without red bulbs there is nothing you can do to turn the tree red. And if you don't have the sockets in the first place it doesn't matter how many red bulbs you've got, the tree still won't light up red. It probably works in exactly the same way for chicken egg yolks and the flesh of farmed salmon, which without a diet of shrimps or commercial canthaxanthin would be an unappetising grey rather than a succulent pink colour.

Carophyll does more than act merely as a dye. If this were all there was to it then all red birds fed on Carophyll would end up the same shade of red, regardless of their species. But they don't. You can feed different species the same amount of Carophyll and watch one turn deep scarlet and the other pink. You can see the same thing in a single species – the redpoll, as its name implies, has a red cap, but its breast is pink and no amount of Carophyll will turn its breast red. This is because there is an interaction even here between the environmental input of carotenes and the bird's genes.[13]

Mendel's Pendulum

The eventual success of Duncker, Gill, Bennett and their followers in producing a red canary mirrors the seemingly endless debate over nature versus nurture. The tragedy of Hans Duncker is that as early as 1922 he recognised the subtle but dynamic interplay between genes and environment in Reich's nightingale-canaries, but failed to apply the same kind of enlightened thinking to anything else. Duncker's explanation for how Reich's canaries acquired their exotic song was an utterly brilliant bit of deduction, but so far in advance of anyone else's thinking that it disappeared without trace, falling unnoticed between the cracks. Perhaps if his paper had been noticed and he had received the appropriate credit, things might have been different. But many years were to pass before the broader implications of this kind of research were recognised. Writing in 1984 Ernst Mayr, the grand old man of ornithology and evolution, summarised the study of song acquisition in birds as follows: 'I do not think that I am exaggerating when I say that research in this area has made a greater contribution towards the invalidation of the rigid polarity of Galton's nature *or* nurture principle than almost any other research.'[14] If only Duncker had realised it, the acquisition of redness in a bird's plumage provided an equally compelling illustration of how inappropriate it was to separate nature from nurture.

But he failed or refused to recognise it. Duncker knew, of course, that colour could have an environmental component because he was well aware of the effect of red peppers on feathers. But creating a red canary by breeding rather than feeding was a matter of personal pride, coupled with an unshakeable belief in the inheritance of colour.

The world of the all-powerful gene was Duncker's world. In a

practical sense it was one that worked well for understanding relatively simple traits, like those so carefully chosen by Mendel in his
peas, and by luck, the colour of ordinary canaries and budgerigars.
Away from the bird rooms and aviaries this narrow type of gene
thinking reinforced an already prevalent belief that what 'is' should
'be'. It also reinforced the notion that differences between human
races, classes and the sexes arise from inherited, inborn distinctions.
As a consequence, theories based exclusively on genetics failed as
miserably with people as they had with the red canary. Moreover,
because the biologists' deterministic theories had lent themselves so
readily to political abuse, genetics became a dirty word and for a long
time after 1945 biological explanations of human nature were taboo.

The taboo lasted thirty years (even longer in Germany) and was
eventually challenged by what many saw as a new form of biological
determinism – the rise of 'sociobiology in the mid-1970s. Tragically,
this name, given innocently enough, was horribly close to the Nazis'
'Sozialbiologie' and triggered a wave of widespread revolt. The sociobiologists' main theme was that genes *do* influence behaviour, but not
to the exclusion of all other influences. In the controversy that subsequently raged within academic and non-academic circles, views became
polarised and it was easy to think that the sociobiologists saw only
genes – a view that Dawkins seemed to confirm in *The Selfish Gene*.
This was wrong, however, as Dawkins was at pains to point out, and on
being asked why sociobiology was so often linked with right-wing
views Dawkins answered, 'Because the opponents of sociobiology are
too stupid to understand the distinction between what one says about
the way the world is, scientifically, and the way it ought to be politically.' This was the root of the problem; concern over the true nature of
human nature – free will versus determinism.[15] The critics of sociobiology were terrified that providing biological or genetic explanations for
behaviour – and human behaviour in particular including war, race
and genocide – exonerated those who had committed and benefited
from monstrous wartime crimes. The critics of sociobiology were

preoccupied with guilt, a legacy of World War II. What they wanted to believe in was free will, free of any genetic influence. Total freedom meant individual responsibility, without which human behaviour, they said, could all too easily be 'explained away', leaving no one morally responsible for his or her actions.

Critics also assumed that science with socially undesirable implications, that is morally or politically bad science, must also be *scientifically* bad. Duncker's science was, for the most part, 'good' in that it was logical and clear-thinking, and his failure to create a red canary can be attributed to two things. First, replacing a yellow canary's genes with those from a red siskin proved far more difficult than he anticipated – especially with the primitive technology and limited genetic knowledge then available. Second, Duncker was so obsessed by genetic effects he never entertained the possibility that red genes even in their rightful owner might require something from the environment (like carotenoids) to make them work.

By the 1990s it was clear that the critics' fear of a new biologically deterministic regime was unfounded. The sociobiologists or behavioural ecologists had won the debate, in part because, with the advent of the human genome project and its increased ability to combat disease, 'genetics' had become a term of promise rather than threat. The genetic pendulum had swung back.[16]

Some people find the idea of being able to partition traits like song or colour in canaries or intelligence in humans into a nature and a nurture component reassuring. But this misses the point. Despite its worthy pedigree, the phrase 'nature versus nurture' which Francis Galton borrowed from Shakespeare's *Tempest* and which Shakespeare borrowed from an earlier writer[17] is utterly outdated. There is no nature-nurture battle. In his book *Eugenics* (2002) David Galton says: 'The nature-nurture antithesis is well and truly dead. It should be consigned to the waste bin together with other moribund ideas'.[18] As the red canary story shows, most traits cannot be divided unequivocally into an either-or situation.

Instead, colour, like so many traits in humans and non-humans alike, arises from the subtle interaction between genes and the environment. The case is most elegantly made by Matt Ridley and encapsulated in the title of his book *Nature via Nurture*.[19]

The idea of nature *versus* nurture is dead, but what is not dead is eugenics. Our enormous strides in understanding the genome and in developing reproductive technologies (including *in vitro* fertilisation and more recently the cloning of Dolly the sheep) means that there are many more eugenic decisions we have to face in future. It has been suggested that these two topics together should be named *reprogenetics*, but as David Galton has pointed out, avoiding the term eugenics obscures the historical perspective: 'The history of the subject is so important because it shows us in no uncertain light the many horrendous pits that we should try never to fall into again. These have been mainly dug by our politicians and regulators.' His view is that if managed properly eugenic technology holds great promise for improving the quality of life.[20]

Postscript

As the 'threat' of genetics and Duncker's politics slowly fade, German academics are gradually starting to address genetic issues in the study of animal and human behaviour, and German bird-fanciers are once again excited by coloured canaries. Not only has the red canary survived a difficult and acrimonious birth, it seems to be here to stay. This might sound obvious, but other once popular canaries, like the London Fancy, are now extinct, and despite years of effort, the particular genome that produced it has proved (so far at least) impossible to re-create. Others, like the lizard, fluttered within a filoplume of extinction, but were brought back from the brink by a subset of dedicated men. As far as the red canary is concerned, past disputes over colour-feeding are almost forgotten and fanciers are looking to the future, striving not only for better, redder, brighter birds, but for new varieties as well. In the quest for a redder canary, breeders who happened upon particular genetic combinations have established an array of wonderful red mutations – variations on a theme – which range from rose-pink to almost purple.[1] What's more, by crossing the true red canary with gold-finches and other native finches they have created ever more colourful birds whose genetic origins lie not in two, but three different species. There is even a small subset of enthusiasts desperately trying to produce blue and black canaries.

What really makes a red bird red remains something of a mystery. Carotenoids are clearly important, but saying that carotenoids make birds red is equivalent to saying that electricity releases the information from a compact disc. To understand fully how a CD works, and why some CDs work in a PC and not a Macintosh (and vice versa), one needs to know something about computer design or to have the help of a computer engineer. To unravel the mystery of colour, evolutionary biologists have recently done the same and combined forces with biochemists. The most prominent of these is the Italian Riccardo Stradi, who also happens to be a bird-fancier and who has made a discovery that may explain the elusive nature of the red canary's colour.[2] The starting point of Stradi's research was the fact, well known to bird keepers, that in captivity most red birds, like house finches, linnets, crossbills and redpolls, lose their redness during the moult and turn yellow, while others like the red siskin are much better at retaining their redness. Stradi recorded the presence and absence of different carotenoids in the feathers of wild caught birds and compared those species that turn yellow in captivity with those that don't. The crucial difference was that red siskins can convert the common yellow dietary carotenoid, lutein, into a red carotenoid and fill their feathers with it. Lutein is superabundant in the diets of all captive finches, but crucially only the red siskin (and one or two other species) has the genetic ability to use it to retain its redness. Were he still alive, Duncker might even have been convinced by Stradi's gene by environment interaction that produces red siskins and red canaries.

One mystery remains. If Duncker successfully transferred the vital genes that confer the physiological trick of turning lutein into red pigment from the red siskin to the canary, why aren't red canaries as red as a red siskin? Part of the answer may lie in a discovery that Duncker himself made during his variegated canary experiments with Reich in 1924: the canary's colour is controlled by a number of different genes. The same is also likely to be true of the red siskin's colours. As geneticists are now well aware, most traits are controlled by several different genes and accordingly their inheritance is more complex and unpredictable than Duncker could have imagined. It is possible that not *all* the red siskin's red genes were transferred to canaries. Another possibility is that even if all the responsible genes do now reside in the red canary, the environment in which they currently find themselves, the canary genome, doesn't allow them to work in the same way as they did in their original 'host' – even in the presence of Carophyll. Studies of genetically engineered mice and other organisms reveal that a foreign gene in a novel genome doesn't always behave the way it 'should'. Despite all the media hype surrounding 'The gene' and the human genome project, there is still a lot we don't know.

The final possibility is that the siskin's red genes are, as Duncker suspected, diluted by the canary's yellow genes – Duncker's red–yellow battle. I sent Riccardo Stradi some feathers from red canaries that had not been colour-fed for him to study their biochemical composition. His results revealed that the red canary has inherited the red siskin's ability to transform yellow lutein into red, but – and this is the crucial part – it still retains the canary genes for changing yellow lutein into yellow – hence the less than blood-red birds. When Duncker planned his experiments he assumed that the siskin's red genes and the canary's yellow genes would lie in the same position on the same chromosome so that when the two species crossed the red genes would oust the yellow ones. Stradi's findings show that this hasn't happened. Although the red canary has acquired the siskin's red genes, it still has the genes that make yellow in full working order. So Duncker, it seems, was right all along.

A Note on Sources

Hans Duncker's private and scientific life has been reconstructed from a variety of sources. These included obituaries (usually all too brief), records kept at the Staatsarchiv in Bremen (including the typescript of the Allies Interview and his 'personal record') and his own publications. A handful of letters (mainly from the 1940s and 1950s) exist and are held in the Staatsarchiv and the Übersee-Museum in Bremen. There were no diaries that I could find and I was unable to locate any members of Duncker's family. Fortunately Duncker was a prolific writer and produced both scientific and popular accounts of his work and the latter often include information of a more personal nature. On the other hand, many of these popular accounts were published in bird-keeping magazines or newspapers (such as *Kanaria*, which came out weekly) but are no longer published and copies from the 1920s and 1930s are now extremely few and far between. I talked to a few people who were contemporaries of Duncker and knew of him, but only one (Professor Gerd von Wahlert) had met Duncker. Throughout the text I have tried to make it clear, usually by identifying a specific reference, where I have evidence for a particular event taking place. Rather than citing every source of information on Duncker here, anyone interested in these should consult Birkhead, Schulze-Hagen and Palfner (2003). Where no specific information exists I have occasionally used my own judgement, based on my own bird-keeping and scientific experience, to infer what might have taken place. I have tried to make it clear where I have done this.

A Chronology of Canaries

Late 1300s – early 1400s: canaries first imported into Spain and Portugal.

1478: Louis XI of France bought large numbers of canaries (Legendre 1955; Yapp 1982).

1534: Conrad Gessner reports seeing canaries in dealers' shops in Europe and that they were expensive.

1544: William Turner refers to canaries in England.

1580: Marcus zum Lamm in Germany illustrates canaries with patches of yellow plumage, marking the start of the domestication process.

1601: Valli da Todi (Italy) publishes a bird-keeping book and starts the shipwreck story of the canary's origin in Europe.

1622: Olina (Italy) publishes the most detailed account of the canary so far.

1657: Johann Walter paints what is probably the first all-yellow canary (see Birkhead, Schulze-Hagen and Kinzelbach 2003).

1705: First edition of Hervieux's monograph on canaries.

Late 1700s – early 1800s: the start of distinct and fancy canary breeds.

1870s: Colour-feeding of canaries on red peppers in England.

1902: First commercial importation of red siskins into Europe.

1914: Bruno Matern (Prussia) has the idea of creating a red canary.

1925: Bruno Matern has created a strain of canaries he calls Rastenburg Reds.

1926: Hans Duncker and Karl Reich make their first red siskin x canary crosses.

1940s: Recognition by C. Bennett in the USA and A. K. Gill in the UK that dietary carotenoids are essential for making red canaries red.

1947: C. B. Bennett's book *A Study on the Nature of the Orange Canary* published.

1955: A. K. Gill's book *New-coloured Canaries* published.

1964: First use of Carophyll to enhance redness in canaries.

Notes

PREFACE

1 Duncker (1922a) describes his meeting with Reich.
2 Genetic imprinting is when the same gene produces a different effect depending on whether it is inherited from the male or female parent; a gene x environment interaction is an instance in which the same gene(s) produces different results in different environments; maternal effects are non-genetic due to differential effects a mother has on her offspring, for example, putting more testosterone into the first laid egg of a clutch.

1. IGNITING THE GENOME

1 The account of the South American siskin catcher is based on my experience with Spanish bird catchers taking small finches using bird lime, a technique that has remained unchanged for a millennium or more.
2 The ancient Chinese kept cage birds, including the nightingale, as did the Egyptians, the Greeks and the Romans (Roberts, 1972).
3 For an account of the development of the Royal Society for the Protection of Birds (RSPB) in the UK see Samstag (1988); for a discussion of the class divisions in animal protection see Ritvo (1987).

2. CATCHING, KEEPING AND STATUS

1 See Parsons (1987), Knolle (1980), Gasser (2001); also visit the Harzer Roller Canary Museum at St Andreasberg, Germany (*www. sankt-andreasberg.delsamson.de* or www.kanarienvogel-museum.de).

2   Information and references regarding Duncker's personal life are in Birkhead, Schulze-Hagen and Palfner (2003).

3   See Nyhart (1995) on Ehlers, and Stein (1988) on Haeckel; Duncker's thesis Duncker (1905a).

4   Allies interview and information in the Staatsarchiv, Bremen; also Birkhead, Schulze-Hagen and Palfner (2003).

5   See Knolle (1980) and Gasser (2001); surprisingly little has been written about the role of canaries in mines. Clarence Royston, a retired miner I spoke to in South Yorkshire who had been part of a rescue team that used canaries to warn them of bad air, told me that it wasn't until 1990 that canaries were replaced by gas detectors.

6   See Aldrovandi (1599), Valli da Todi (1601), Olina (1622) and Solinas (2000).

7   See Markham (1621).

8   Vergil quoted in MacPherson (1897); see also Cox (1677), Fortin (1660), Blagrave (1675) and Bub (1991) on bird catching.

9   See Duncker (1905b) for his study on migration and Berthold (1993) for a modern account. Natalino Fenech (1992) provides a horrifying but highly readable account of contemporary hunting and trapping on Malta (see also Fenech 1997).

10 See Dixon (1851); see also Carluccio and Carluccio (1997) *Complete Italian Food*, under Beccafico (Blackcap *Sylvia atricapilla*): 'Various small birds belong to this family . . . Their meat is especially flavoursome when they feed on figs . . . They are quite rare birds and because of this their hunting is usually illegal. The majority of those sold in specialist shops come from abroad so that the consciences of the Italians who enjoy eating them are not troubled.' Ortolans: the multiple gustatory orgasm is not solely a woman's pleasure. As former French President Mitterrand was dying of cancer one of his final wishes was to indulge himself once more in an unlawful passion – consuming ortolans – long since protected by law. In a masonic ritual Mitterrand, surrounded by twenty of his closest friends, places a napkin over his head to better capture the redolent vapours of the little birds he is about to eat. Then, 'with the gesture of an omnipotent emperor, he plucks a tiny bird from a steaming bowl and noisily

sucks in flesh, bones and blood before glancing around the room "dizzy with contentment, his eyes sparkling" ' (from *Le Dernier Mitterrand* by Georges-Marc Benamou). Mitterrand's lifelong love of little birds was driven by more than oral pleasure; he was also motivated by the status forbidden fruits engender. As, indeed, are others and as I was writing this I heard of an internationally renowned ornithologist who had eaten a bowl of little birds – beaks and all – in the same unquenchable quest for status.

11 Duncker (1905b); for trapping techniques see MacPherson (1897) or, more easily but less romantically, Bub (1991); for the location of *roccoli* see Tappen (1959) and Calegari et al. (1985).

12 See Holzinger (1987).

13 See Belon (1555), Olina (1622), Ray (1678), Raven (1947), Birkhead (2010).

14 Of all the things bird trappers (used to) do to birds, blinding is the one that evokes the strongest feelings of disgust. Blinded decoys were popular because, being less easily distracted, they sang more and were more effective. The earliest record of blinding I found is in the German bird-catching book by Aitinger (1626), but it certainly occurred before then. Albin (1731–38) describes the process: ' 'Tis a cuftom among the bird-men, when they want to learn the chaffinch a song, to blind him when he is about three or four months old – which is done by clofing his eyes with a wier made almost red-hot, because as they say, he will be more attentive and learn the better; but I am sure it would be much better to never confine them in cages than purchase their harmony by such ufage.'

Charles Waterton, the eccentric, pioneering English bird conservationist, wrote an impassioned account of a blinded chaffinch he saw during a visit to what is now Belgium in the 1790s: '. . . you see it outside the window, a lonely prisoner in a wooden cage, which is scarcely large enough to allow it to turn round upon its perch. It no longer enjoys the light of day. Its eyes have been seared with a red hot iron in order to increase its powers of song . . . Poor chaffinches, poor choristers, poor little sufferers!' Waterton (1870)

Before blinding their birds, keepers trained them to find their food and water in the dark, so that once permanently deprived of

sight they could still look after themselves. Aitinger's engraving illustrates the method Waterton describes, but other methods, no less
disgusting, were also used. These included pouring drops of caustic
soda into the eyes – trappers referred to these birds as 'bubble eyes',
for reasons that are not hard to imagine. Others simply destroyed the
eye with a knife. In England, Fred Speakman described how most
people blinded birds with a hot pin, but because this caused an
unsightly white spot in the eye some used a hypodermic syringe
loaded with black ink (Curtis and Speakman, 1960). Occasionally
keepers actually sewed the birds' eyelids together. In the Netherlands
a more sophisticated, but no less cruel, method was used: the hot wire
was used to seal the eyelids closed, but the eye itself was undamaged
and the process could be reversed (Hoos, 1937).

What is surprising is how long blinding continued when it could so
easily have been replaced by the simple expediency of covering the birds'
cages with a cloth or, as the Maltese do with decoy turtledoves, giving
them a hood. Mercifully, blinding is now scarce. The Austrians banned it
in 1870, snidely remarking that the Italians failed to do so for over another
hundred years (it was banned in Italy in 1977) (Kuthy, 1993); in Britain
blinding was banned with the Wild Animals in Captivity Act (Anon.,
1905) and it was made illegal in Spain in the 1980s.

> Does it matter? – losing your sight? . . .
> There's such splendid work for the blind;
> And people will always be kind,
> As you sit on the terrace remembering
> And turning your face to the light.
>
> SIEGFRIED SASSOON

15 See Deelder (1951); for more information on *vinkenbaans* see Hoos
(1937) and Matthey's book *Vincken moeten vincken locken* (the title is
a quote from a seventeenth-century Dutch poem by Jacob Cats,
which means: To lure finches you need finches. The subtitle, *Vijf
eeuwen . . .*, means: The catching of songbirds and quail in Holland
during the last five centuries).

16 See Vale (1974).

17 See Wright's (1994) *Moral Animal* and Miller's (2000) *Mating Mind.*

18 Darwin (1871).

19 See Miller (2000), Dawkins's (1982) *Extended Phenotype.*

20 Jongh (1968) describes the remarkable erotic subtext of many apparently innocent Dutch paintings. See also Bewick (1862, p. 55) who says, 'I was also entertained with the curious characters who resorted to his house – these were mostly Bird catchers and Bird dealers, with whose narratives respecting their pursuits, I listened to with some interest while they were enjoying themselves over a Tankard of Beer. Ned was almost constantly busied in rearing a very numerous brood of Canaries, which he sold to a bird Merchant, who travelled with them at set times to Edinburgh, Glasgow &c for sale. . . . I also got fully into a knowledge of the misguided ways, which too many fellows pursued . . . a chap who was chopped down in his youthful prime solely by his connecting himself with the bad Women of the Town & becoming perfectly tainted by his intercourse with them . . .'

21 Oxford English Dictionary.

22 Bird as penis. The most graphic example of this is a French or Flemish illustration from the first half of the seventeenth century, *La chasse à la pipée,* reproduced in Ignaz Matthey's book (2002, p. 199) on bird catching. It shows seven winged penises around a *passera,* while two bird catchers look on from behind a bush.

23 MacPherson (1897). That part of the Reverend MacPherson's motivation for writing his *History of Fowling* may have sprung from his own dual fascination for bird catching and sex is suggested by a particularly telling phrase in the preface of his book. At the invitation of count Camozzi Vertoa of Bergamo, MacPherson visited Italy to observe a *roccolo* and obviously had a wonderful time there. The count 'gave me a very happy impression of Italian hospitality, which was more than confirmed by my intercourse with the various members of his family'.

24 Information on Duncker's career is from the Staatsarchiv, Allies Interview and obituaries; see also Birkhead, Schulze-Hagen and Palfiner (2003).

3. THE MUSIC OF NATURE

1   See Albin (1737), Roberts (1972).

2   See Buffon (1793), Bechstein (1795); Isaak Walton is quoted in Armstrong (1958), Mayhew (1861).

3   See Arnault de Nobleville (1751), Andrewes (1830), Pernau, cited in Stresemann (1947) Birkhead and Charmantier (2013).

4   See Stresemann (1947, 1975), Schlenker (1982) and Thielcke (1988). No portrait of Pernau exists. He apparently taught canaries to sing a nightingale song but via a linnet tutor, describing in his book (1707) how he trained young linnets to mimic nightingale song and then used these in turn to tutor young canaries!

5   Barrington (1773).

6   Syme (1823).

7   See Knolle (1980), Gasser (2001).

8   See Hervieux (1708) and Chapter 5 for more on the remarkable Hervieux.

9   Sadly, this story is almost certainly apocryphal. It comes from Legendre (1955a) but I have been unable to verify anything Legendre says about the early history of the canary. After presenting this story of Elizabeth I he goes on to say that a yellow bird appeared among the queen's canaries. This was so striking that Shakespeare wrote about the 'miraculous' transformation of the green bird to a golden yellow one in one of his poems, attributing the transformation to 'the gaze of a queen more empowered to produce gold than the Atlantic sun'. I checked this with Dr Mick Hattaway, a Shakespeare expert, and he didn't recognise it, nor does Harting (1871) mention canaries in his analysis of Shakespeare's birds.

10 Gessner (1555); Willughby in Ray (1678).

11 The history of the canary comes from various sources, including Galloway (1911), Wetmore (1938), Speicher (1976), Parsons (1987, 1989).

12 See Pernau in Streseman (1947), Hervieux (1708), Buffon (1793) – M. Hébert, the tax collector, was one of Buffon's many informants.

13 See Barrington (1773), Knolle (1980), Gasser (2001).

14 St Andreasberg Museum. Although St Andreasberg still promotes

itself as a centre of canary breeding, when I visited it in 2001 I found only a single old man who still bred canaries there.

15 See Knolle (1980), Banks (1976), Wetmore (1938).

16 Gleich et al.

17 Parsons (1987, 1989).

18 For information on Reich see Ringleben (1955), and also the Staatsarchiv, Bremen.

19 Duncker wrote several articles about Reich's nightingale-canaries (see Birkhead, Schulze-Hagen and Palfner (2003)), but the interview conducted by Grenze (1938) in the weekly canary newspaper *Kanaria* is the most readable.

20 The term 'stopping' refers either to stopping the light or stopping birds from singing by eliminating the light. Stopping is mentioned by many early authors including Aldrovandi (1599), Aitinger (1626), Anon. (1728), Damsté (1947), Hoos (1937). Aldrovandi says, 'About the beginning of May . . . Accustom them to darkness over ten days until completely dark that hath not the least chink to let light in. . . . During all the time of their imprisonment in this dungeon, nobody must come in there but their keeper, and he with a candle once in three days to give fresh water and meat, and to cleane their cages . . . In this manner they must be kept till about the tenth of August.'

21 Copeland et al. (1988) provide a nice account of Reich's pioneering role in birdsong recording.

22 Darwin (1871).

23 See Broeckhoven (1969), Kragenow (1981), Bergmann (1993); see also the Avibo website: info @ avibo.be.

24 For singing behaviour of the chaffinch in the wild see Thorpe (1961); in contests see Santens (1995); Filip Santens, pers. comm.

25 See Rettich (1896), Curtis and Speakman (1960).

26 Sheldon and Burke (1994).

27 Santens (1995).

28 Blakston et al. (1870, p. 295).

4. MUSIC IN THE BRAIN

1   Darwin (1871). For the complex but fascinating link between the regions of our brains that are stimulated by language, music and bird-song see Gray et al. (2001).

2   See Kroodsma (1976).

3   Sexy syllables were discovered by Eric-Marie Vallet and co-workers (Vallet et al., 1998); see also Güttinger (1985), Mundinger (1995) and Leitner et al. (2001) who explore the effect of domestication on canary song.

4   See King-Hele (1999).

5   Barrington (1773).

6   Gaunt and Wells (1973).

7   See Herrick and Harris (1957), Gray et al. (2001), Specter (2001), Birkhead et al. (2014: 211–13).

8   From Bowler (1983, pp. 265–268) on Lamarckism, and J. W. (Pim) Arntzen, pers. comm.

9   See Hervieux (1708). The first edition of Hervieux's book was said to have been published in 1705 but apparently no one ever saw it. The earliest extant edition is 1708.

10 Barrington (1773).

11 See Thorpe (1961).

12 See (Anon., 1714) *The Bird Fancyer's Delight* and Godman (1955), Hammersely (1717), Thorpe (1955).

13 See Wells (1993).

14 See Thorpe (1955).

15 For information on the serinette I thank Françoise Dussor; see also Zeraschi (1980), Baines (1992).

16 See Bechstein (1795) and Newton (1972).

17 See Schulze-Hagen and Geus (2000) for Wolf's meeting with Darwin; see Nicolai (1956, 1993) and Güttinger et al. (2002) for bullfinches learning German folk tunes.

18 The bullfinch is a mystery: its biology is unlike any other finch, as is its singing behaviour. It does sing, but quietly and not to advertise ownership of a territory for it doesn't defend one. Rather, the male bullfinch sings only for the benefit of its partner and its song is always

accompanied by a beautiful tail-twisting display, making them even more endearing to their partner and human owners. I have listened to old recordings of 'piping bullfinches' and felt that the novelty of a bird whistling two German folk tunes would soon wear off, but in fact bullfinches sing so infrequently there was little chance of their owners getting fed up with them. Albin (1737) wrote, 'The expense and trouble required to bring them to this state of perfection cause them to be very dear; and they are seldom found in this country, except among the rich.'

19   Nicolai (1956, 1993).

20   See Syme (1823).

21   Duncker was so excited about understanding Reich's nightingale-canaries that he wrote articles for no fewer than four different magazines: three popular ones (*Kosmos, Kanaria* and *Gefiederte Welt*) and one scientific (*Journal für Ornithologie*) – the details of all Duncker's publications are in Birkhead, Schulze-Hagen and Palfner (2003). The standards set for singing canaries by the Deutsche Einheitsskala in 1922 have remained almost unchanged to the present time.

22   See Huxley (1942); P. Marler, pers. comm.

5. THE VARIEGATION ENIGMA

1   These are: Durham and Marryat (1908) and Davenport (1908).

2   See Duncker (1928) *Genetik der Kanarienvögeln*. When I refer to 'yellow genes' or, later, 'red genes' – this is shorthand for genes that code for particular proteins that result in one or another particular colour.

3   Burleigh (2000).

4   Hervieux's wonderful little book went through no fewer than ten French (1705–1802), eight German (1712–71), one Italian (1724), one Dutch (1712) and one English (1719) editions (R. Schlenker, pers. comm.).

5   Information on the Princesse de Condé and her relationship with Hervieux comes from Milon (1956), Fisher (1966), Schiavone (1978), Kroll (1998) and Legendre (1955a, b) whose information, regrettably, cannot be trusted. The Grand Condé was keen on acquiring

interesting birds for his aviaries and there is a record from 1682 of him negotiating to purchase an extraordinarily beautiful bird, a 'coq de Virginie' just arrived from the Indies (in fact, a North American northern cardinal), a flame-red bird with a huge crest and a voice somewhat like a nightingale (Robbins, 2002). I have been unable to trace a portrait of Hervieux. There has been some confusion in the bird-keeping literature about the true identity of Madame la Princesse, for example, Galloway (in his chapter in Robson and Lewer (1911, p. 11)) says that he had been informed by a leading Dutch canary man, C. L. W. Noorduijn of Groningen, that Hervieux was 'Inspector of Canary-breeding to the Duchesse de Berry' and Gill (1955, p. 8) says something similar, but this is wrong.

6 Pernau (1702) cited in Stresemann (1947).

7 See Valli da Todi (1601), Olina (1622), Barbagli and Violani (1997). Shipwrecks were a popular way for novel species to appear on distant shores: the much-prized Spanish merino sheep was said to have arrived in England from Spanish Armada ships wrecked on the English coast (Wood and Orel, 2001).

8 See Anon. (1873).

9 See Barbagli and Violani (1997).

10 See Valli da Todi (1601), Solinas (2000); for Olina as a plagiarist see Schwerdt (1928). Interestingly, it turns out that Valli da Todi was a plagiarist too: he lifted most of his material on the canary (but not, it seems, the Elba story) from the book by Manzini (1575).

11 This is from Legendre (1955a, b) again, so it may not be true. Indeed, on reading Major's (1872) account of Bethencourt's voyage I found no mention of canaries.

12 See Parsons (1987) and for Louis XI see Yapp (1981, 1982). Many articles of trade were subject to strict regulations; the monopoly on canaries was simply one of many: the Spanish tried to regulate trade in other livestock too, like their merino sheep (Wood and Orel, 2001). For the most recent estimates of the total populations of birds in Europe see *BirdLife* (2000).

13 See Evans (1903), Gessner (1555).

14 See Hervieux (1719) and Ray (1678). In the English translation of

Hervieux (1719) the foreign canary traders are referred to as German but others (e.g. Robbins, 2002) refer to them as Swiss. The differences exist because national boundaries have changed over time.

15 See Andrewes (1825).

16 See Ritvo (1987) and Darwin (1859).

17 See Galton (1865) and Diamond (1997).

18 See Kroll (1998).

19 See Roberts (1972).

20 See Ray (1678).

21 See Durham (1926), Nicolai (1956), Goodwin (1965). I went to see some wild-caught canaries but they were so wild that it was impossible to get a good look at them.

22 Hervieux (1719).

23 See Ritvo (1987), Zeuner (1963) and Sossinka (1982).

24 Being bored makes you boring: see Specter (2001).

25 Specter (2001).

26 Numbers of canaries in London – see Anon. (1742); I wonder how he arrived at this figure.

27 See Albin (1737); for background information see Jackson (1985).

28 See Stresemann (1975).

29 See Haffer (2000).

30 See Price (2002) on the use of information from cage birds to understand evolutionary processes relating to speciation.

31 See Stresemann (1923a) and also Hackethal (1990) on Röting.

32 Change of colour from green to yellow is described in Stresemann (1923b) but also see Newton (1896), Cox (1677), Blagrave (1675) and Birkhead, Schulze-Hagen and Palfner (2003). For Pepys see Godman (1955).

33 See Kinzelbach and Holzinger (2001) for Lamm's bird volumes.

34 See Harley (1982) and Ball (2001).

35 See Anon. (1994). It is just possible that pure-yellow canaries existed much earlier still. The biological historian R. Kinzelbach found a coloured illustration in an Italian school book produced in 1490, of two boys each with a pure-yellow bird, which may well be canaries (see Birkhead, Schulze-Hagen and Kinzelbach (2003)).

36 Lander and Partridge (1998) is the most up-to-date book on British bird keeping; the numbers of birds in Europe is from *BirdLife* (2000).

6. DOMESTIC LIFE AND DEATH

1   See Janet Browne's (1995) *Charles Darwin: Voyaging.*
2   Blakston et al. (1870).
3   See Albin (1737) and Hope 1762 (but note that the latter was based on Anon., 1742).
4   See Evans (1996a, b) for the history of the lizard canary.
5   The early history of canary societies is poorly documented, but see Anon. (1846), which was probably written by Harrison Weir, a co-founder of the *Illustrated London News*, one of Darwin's correspondents, and a top canary judge and illustrator. See also Hope (1762) and Syme (1823).
6   See Ritvo (1987).
7   There is confusion regarding books by 'Andrew', partly because there were several different editions (some published anonymously), but also because Andrew is sometimes referred to as Andrewes. This one is: Andrewes (1830), *The Bird Keeper's Guide and Companion.*
8   Haeckel (1905).
9   See Browne (1995), but also Secord (1981, 1985).
10 Browne (1995).
11 Mutations. In most instances mutations have little visible effect. All genes come in pairs, referred to as alleles, and if one allele of the pair is defective the other usually compensates for it and everything looks normal. But if both alleles are defective the effect is often noticeable. This is what happens with albinos: albinism is a recessive mutation and thousands of people carry the gene for albinism – but because they carry just a single allele it is invisible. Only individuals who possess *both* alleles coding for albinism lack pigment and are referred to as albinos. The most likely way albino individuals arise is if two fully pigmented individuals, each with one albino allele, produce an offspring blessed with both albinism alleles. Among humans the incidence of albinism is very low, about one in 20,000 in Europe, but obviously if albinistic individuals reproduce with their relatives, the

chance of having albino offspring is greatly increased. Put another way, inbreeding amplifies the incidence of albinism, because the chances of partners carrying the same allele are increased. We are not just talking about skin pigmentation here, most genetically determined traits, including those which control the colours in birds' feathers, are more likely to be expressed in the offspring of closely related parents – that is, when inbreeding occurs. The fact that white-flecked birds became more common in the Ukraine following the Chernobyl nuclear accident provides a compelling, if somewhat chilling, reminder that radiation damages chromosomes and creates mutations. For the effects of radiation from Chernobyl affecting plumage colour see Ellegren et al. (1997).

12   T. H. Morgan in 1908 was the first to use X-rays to create mutations in fruit flies, but it was his brilliant student, H. J. Muller, who showed in the mid-1920s that that radiation damaged the chromosomes and caused (mainly deleterious) mutations that could be inherited (Allen, 1978). For Duncker's warning to Meyer about radiation, see Duncker (1930).

13   For more information on the use of back-crossing to capture particular mutations or traits see Henig (2000) – a popular account of Mendelism – and Wood and Orel (2001). The quote by Pernau (1702) cited at the beginning of this chapter is from Stresemann (1947). I subsequently found indirect evidence of an even earlier example of back-crossing: in Marcus zum Lamm's 'encyclopaedia' of c.1580 he refers to an established strain of white goldfinches (see Kinzelbach and Holzinger, 2001).

14   Information from Keynes (2001) and R. Keynes, pers. comm.

15   Apart from having a cage bird in the room of an ailing child to take the sickness away, small birds (or bits of them) were also used directly as medicines. For example, a soup made from hoopoes would cure headaches and stimulate milk production in nursing mothers; and verrucas – which in Austria are called chicken-eyes – could be cured by squeezing the eye of a hoopoe on to them. The heart of a swallow was said to improve the memory. Crossbills apparently had several outstanding properties, and in some areas of Germany and Austria it

was said that placing a caged crossbill under the wedding bed would ensure the conception of a son. Crossbill claws worn as a necklace would stop toothache and simply having a crossbill in the house would prevent a lightning strike. Crossbill water, whatever that is, was especially good for madness and teething children (Gattiker and Gattiker, 1989; Kuthy, 1993).

16   Charles Dickens's *The Old Curiosity Shop*.

17   Darwin got the idea of feather-footed canaries from a book by Brent, *The Canary, British Finches, and some other Birds* (1864). Indeed, he underlined this bit in his copy. Brent in turn obtained the term from the English edition of Hervieux (1719) in which the translator for some inexplicable reason mistranslated the term 'duvet' (meaning down feathers) as 'rough-footed' (see Robson and Lewer, 1911, p. 26), which Brent then proceeded to call feather-footed and feather-legged. Darwin simply assumed Brent to be correct. However, Brent does say (p. 22), 'The rough-footed or feather-legged Canaries now seem to be very scarce, if the breed is not altogether lost, as I do not remember having seen but one, and that many years back.' Brent (1822–67) was a fancier of pigeons, fowl, canaries, British finches and mules. Darwin met him in the 1850s through his interest in pigeons and described him as 'a very queer fish'. A recently discovered photograph of him shows Brent to have been a very small man. Brent wrote regularly for the *Cottage Gardener* to which Darwin subscribed, and a series of articles on canaries and finches published during the 1850s formed the basis for his book.

18   Blakston et al. (1870); Roberts (1972); Haffer (2001).

19   For an account of the rediscovery of Mendel see Provine (1971).

20   Lipset (1980).

21   Gillham (2001) provides a good overview of the dispute between the Mendelians and biometricians; see also Provine (1971).

22   See Pauly (2000).

23   Davenport (1908).

24   Galloway (1909).

25   Follow the fight in Davenport (1910), Galloway (1910) and Heron (1910).

26 For a historical account of evolutionary ideas in Germany see Mayr (1980).

27 Provine (1971) provides an excellent history of genetics; see also Mayr (1980).

7. MIXED BLESSINGS

1 See Blakston et al. (1870).

2 See Bemrose and Orme (1873).

3 See Barnesby (1877).

4 The practice of producing artificially coloured parrots in this way was known long before Bougainville's voyage: it is described in Willughby's *Ornithology* (Ray, 1678) and was probably discovered soon after Europeans discovered South America in 1500. Ray writes (Ray, 1678, p. 110), 'Parrots are made of several colours by the Tapuyae [a group of people of Brazil], by plucking them when they are young, and then staining their skins with diverse colours. These the Portuguese call counterfeit Parrots. Which thing if it be true (for to me indeed it seems not probable) it is to no purpose to distinguish parrots by the diversity of their colour, sith therein they may vary infinitely.' See also Robbins (2002).

5 After World War II the new AZ was Vereinigung für Artenschutz, Vogelhaltung und Vogelzucht (from Vins, 1993, p. 22). The AZ continues to this day – its official magazine is *AZ-Nachrichten*.

6 Information on Cremer comes mainly from Duncker (1927a), Anon. (1928), which was probably written by Duncker, and Duncker's obituary of Cremer (1938a).

7 See Blakston et al. (1870).

8 Red siskin information comes from Coats and Phelps (1985 and references therein), Collar et al. (1992), Durham (1926).

9 A handful of extraordinarily elusive mules and hybrids continue to fire the imaginations of European bird keepers. For example, in 2001 a hybrid crossbill x bullfinch changed hands for (allegedly) £2500 – presumably on the assumption that it would be a winner at the British National Cage Bird Show in December 2002. It wasn't – apparently because it was stolen before the show. Ever since Hervieux

indicated the existence of chaffinch x canary mules and yellowhammer x canary mules (see Robson and Lewer, 1911, p. 22), bird keepers have dreamed about them. Remarkably, chaffinches paired to hen canaries have very, very occasionally produced fertile eggs, but no surviving offspring (T. Roberts, pers. comm.). Pairings between the yellowhammer x canary have never even got this far, which isn't surprising given their very distant relationship: the yellowhammer is a bunting and the canary a finch. But because they are both yellow, breeders often assumed they were closely related and should actually want to hybridise!

10   See Barnesby (1877).

11   Bechstein's fantasy: 'I possess also a male nightingale, which has long lived and sung in the same cage as a female canary. They were both so ardent last spring, as to pair in my presence, but the eggs proved unfruitful. Should the same thing occur again, I will put the eggs under some other birds' (Bechstein, 1800 edition – but not in first edition of 1795).

12   Albin (1738–volume 3). Interestingly, a melanistic goldfinch whose description matches Albin's painting almost completely was given by MacPherson (1880).

13   Bechstein (1795 editions and following: see Schlenker, 1994) refers to bullfinch x canary hybrids but is confusing because (under Canary) he refers to those bred by both himself and D. Jassop of Frankfurt as being the offspring of a male bullfinch and hen canary, but elsewhere (under Bullfinch) he says that such hybrids are produced by pairing a very ardent male canary to a hen bullfinch. He also says that this hybrid is very rare. The only earlier record I can find is in Anon. (1772) which refers to one arising from a young canary female and an old sexually very active bullfinch male. In view of the fact that virtually all subsequent successes have been achieved by using *female* bullfinches (see Gray, 1958, p. 270) these records should be treated cautiously.

14   Houlton (c.1920).

15   Birkhead (2000); Birkhead and van Balen (2007), Birkhead et al. 2008a).

16   See Galloway (1911).

17 Hervieux (1708); Buffon (1793); Bechstein (1795); Page (1914).

18 Buffon (1793).

19 Potts and Short (1999); Ritvo (1987).

20 Buffon (1793).

21 Ritvo (1987); Burkhardt (1979); Müller-Hill (1988).

22 Müller-Hill (1988, p. 81).

23 Barrett et al. (1987); see also Mayr (1982).

24 Arnaiz-Villena et al (1999).

25 Durham (1926).

26 Fumihito et al. (1996).

27 On the fertility of hybrid birds: see Hervieux (1708), Buffon (1793), Cookson (1840), Darwin (1871); confirmed by later studies, e.g. Promptova (1928).

28 Darwin's *Domestication* (1868) – but as we've seen with the feather-footed canary (above, Chapter 6, note 17), Brent wasn't as reliable as Darwin thought.

29 See Grant and Grant (1992): somewhat uncritical lists of birds that have hybridised are given in Page (1914) and Gray (1958). See also Arnold (1997) for the role of hybridisation in speciation.

30 Buffon (1793); Galloway (1909).

31 Durham and Marryat (1908).

32 Haldane (1922); Galloway (1909); and for an ingenious but almost impenetrable explanation of Haldane's rule see Orr (1997).

33 See Dams (1925, 1926), Gill (1955).

8. FUGITIVE RED

1 See Duncker (1927b), Gill (1955).

2 Birkhead et al. (2008b).

3 Duncker (1929a). Breeding together these different mutants and establishing their genetic constitution and hence the type of offspring they would produce was the foundation of Duncker's success. This solid work still forms the basis for all budgerigar genetics and his breeding expectations are still used today. However, Duncker's notion that the different budgerigar colours were controlled by just three factors (F, O and B – Duncker's so-called FOB theory, in which F =

Fat factor and is responsible for yellow colour, O = Oxydase factor, responsible for blue colour, and B = Brown factor responsible for darkening the colour [see Watmough (1935) for details], which was supported and independently verified by Hans Steiner, Professor of Zoology at the University of Zurich) was later criticised by Crew and Lamy (1934, 1935). Their point was that it was much more likely that several genes were involved in each biochemical pathway resulting in a change in colour. The criticism is a subtle one – splitting hairs – said one geneticist I spoke to (J. Slate pers. comm.), but it must have damaged Duncker's reputation and his pride.

4   The German Budgerigar Society was founded in 1926 by Cremer and Duncker, together with Wilhelm Schinke (a protestant clergyman of Hordorf), Herr Aumüller (Delitzsch), Albert Krabbe (Anklan) and Ado Mertes (Mainz). The Nazis banned the society in 1934 and it restarted only after the war in 1948 (Vins, 1993). As a measure of Cremer's contribution to the AZ, in 1959 the society created the Consul Cremer Prize for outstanding achievements in aviculture.

5   See Haffer's (2000) biography of Stresemann.

6   Duncker (1927b).

7   Eugenics in Germany: Stein (1988); Müller-Hill (1988). See also Galton's (1869) *Hereditary Genius*, D. Galton (2002).

8   See Pauly (2000).

9   Davenport (1908).

10   Stein (1988); Müller-Hill (1988); Gillham (2001) provides a useful overview of Galton and eugenics in the UK and elsewhere.

11   See Gill (1955), who in the 1920s was the UK expert on white canaries. See also Albin (1731–38), Stresemann (1923b) and Anon. (1994).

12   Duncker (1927b).

13   *Gefiederte Welt* is the oldest cage bird journal in the world. First published in 1872, it was the brainchild of bird enthusiast and journalist Karl Russ (1833–99). Russ, who edited the journal from its inception in 1872 until he died in 1899, was well qualified, for he had previously completed a doctoral thesis on crossbills. He wrote several books on cage birds, some of which were translated into English. The second editor was Karl Neunzig (1864–1944) who had been trained at the

Berlin Academy of Art and who painted the red siskin and mule pictures for Duncker's (1927b) paper in *Vögel ferner Länder*. After they came to power in 1933 the Nazis harassed Neunzig because his great-grandmother was Jewish and in 1938 they forced him to resign as editor. Neunzig nonetheless continued, surreptitiously, to provide illustrations for *Gefiederte Welt* during the Nazi regime under a pseudonym: Archiv Creutz. He died in 1944. The third editor, from 1938, was Joachim Steinbacher (born 1911), a student of Erwin Stresemann and curator of birds at the Senckenberg-Museum in Frankfurt. Steinbacher and Duncker corresponded in the 1950s (see Chapter 10) and, remarkably, Steinbacher was still an honorary co-editor of *Gefiederte Welt* in 2003 (see also Hinkelmann, 2001). The British equivalent, *Cage and Aviary Birds*, started in 1902 and, like *Gefiederte Welt*, is still going.

14 See Gaddis (1955) and Stong (1957). Stroud became an expert on cage-bird diseases and his book on this topic is still in print.

9. NOT BY GENES ALONE

1 Information on A. K. Gill's role in creating a red canary comes from Page (1957) and from Koudis (2000). The society Gill started in 1934 went through several name changes, beginning as the Red Canary Movement, then in 1938 changing to the Canary Colour Research Association and in 1947 to the Canary Colour Breeders Association.

2 See Bennett (1947). Charles Bennett remains virtually unknown: he worked in the laboratory of Jacques Loeb, one of the best-known American biologists at the beginning of the twentieth century. I was unable to find a photograph of Bennett.

3 See Dams (1925, 1926).

4 See [British] Budgerigar Society Yearbook (Anon., 1929). A record of Duncker's illness is in the Staatsarchiv, Bremen.

5 Alfred Kühn, a student of Weismann, had a special interest in genetics and developmental physiology, clearly admired Duncker's work and later (in 1937) became second director at the Kaiser Wilhelm Institute in Berlin. The letter from Jentsch asking that Duncker be promoted and the reply to Kühn's letter are in the Bremen Staatsarchiv.

6   Duncker (1929c).

7   Gill was Britain's white canary expert.

8   It later became clear that recessive white canaries did carry genes for yellow plumage after all (see Dean, 1963).

9   See Gill (1955).

10 The 1940s saw a flurry of books on coloured canaries published in the USA. Bennett's (1947) is the best, although it is a rather rambling account. The books by Osman (1948) and Armitage (1947) books are frustratingly imprecise.

11 A recent review of the likelihood of different bird species producing fertile hybrid offspring confirms that Lewis's results were extraordinary (Price and Bouvier, 2002): the canary and the fire finch diverged about 12 million years ago (T. Price, pers. comm.).

12 See Bennett (1947, p. 44).

13 See Keeler (1893a, b).

14 See Brockman and Völker (1934).

15 Bennett (1947).

16 ' "That's the egg of Columbus!" is used in German to describe a simple, unexpected and ingenious solution to a difficult problem. I think there is more than one story behind it. The legend I remember goes that Christopher Columbus was having a meal, I think at the Spanish court, and somebody asked if anybody could make a boiled egg stand upright on the table. Nobody found the solution until Columbus took the egg and simply smashed its bottom on the table so that it stood upright of course! (I wonder if this is a typical German way to sort out things and if that's the reason why the legend is so popular there!)' G. Palfner, pers. comm.

17 Gene x environment interactions have been known for a long time. The Austrian sheep breeder Baron J. M. Ehrenfels, whose views were known to Cyrill Napp, Abbott at Mendel's monastery, wrote in 1831, 'Climate, nutrition and generation remain the levers of Nature in the formation of matter. In the interaction of these three potentials, generation, the genetic force is the most powerful' (cited in Wood and Orel, 2001). T. H. Morgan, working in the early 1900s, wrote about gene x environment interactions (see Allen, 1978). The most

striking example concerns human height – a trait known to be heritable since Galton's day. The Japanese were traditionally renowned for their diminutive stature, but following an increase in the amount of protein in their diet after World War II, their height increased by several centimetres in a single generation.

18   Robson and Lewer (1870); Robson (1918).

19   For house finches on Hawaii see Grinnell (1911) and for canaries on Midway Atoll see Munro (1944). Intriguingly, the canaries on Midway later turned yellow again (Wetmore, 1938) and have remained yellow ever since, probably because huge numbers of alien plant species have been introduced, increasing the availability of carotenoids (Beth Flint, pers. comm.).

20   Hill (2002).

21   Burleigh (2000).

22   Walter (1990) and pers. comm.

23   Meyer and Duncker (1933) on 'unworthy life'. After obtaining a copy of this I gave it to Gotz Palfner, who helped translate much material for me. I was amazed to learn from him that this type of literature is still suppressed in Germany.

24   See Gillham (2001).

25   In a wave of enthusiasm or opportunism, Duncker and another teacher edited a book on new ways of teaching biology consistent with Nazi ideology (Duncker and Lange, 1934).

26   Ball (2000).

27   Duncker (1938b). The standardisation of colour was a long-standing problem. The artist and bird-fancier Patrick Syme attempted to resolve it in the 1820s, with a similar system of reference colours, which Darwin took with him on his *Beagle* voyage and which later his beloved daughter Annie delighted in using (Keynes, 2001).

28   See Duncker's obituary of Cremer (Duncker, 1938a).

29   This is from video footage of the destruction of Bremen (Temmen Video 2001) and first-hand accounts (Garthus, 1977).

30   Haffer (2000).

31   Information on Reich is from the Bremen Staatsarchiv, and the AZ membership lists in post-war issues of the AZ journal.

32   The information on these pages comes mainly from Duncker's Allies Interview held in the Staatsarchiv, Bremen. I also used Garthus (1977) and Napoli (1949). Duncker's offer in 1933 to become director of the Kaiser Wilhelm Institute for Biology at Berlin-Dahlem occurred in response to the vacancy created by the death of the first director, Carl Correns (one of the three researchers who rediscovered Mendel – see Chapter 6). This invitation was a monumental honour, especially if it was motivated purely by science, since the Kaiser Wilhelm Institutes (renamed in 1948 as Max Planck Institutes) were the premier independent research institutes in Germany. Duncker said he declined the invitation because it was conditional on his joining the Nazi party. Ironically, at other Kaiser Wilhelm Institutes, highly rated researchers who refused to join the Nazi party continued to be well funded and unharassed during the war (see Deichmann, 1996), although Duncker could not have known this at the time of his invitation. The post went instead to the botanist Fritz von Wettstein, considered to be the best plant geneticist in Germany. His research on plant development, subspecies formation and mutation later became well known. Wettstein never joined the National Socialist party. I discussed Duncker's refusal to accept the KW position with Karl Schulze-Hagen, and we independently came to the conclusion that Duncker probably wasn't well enough trained to make the most of it: Duncker himself was probably smart enough to realise this and consequently declined the offer (see Birkhead, Schulze-Hagen and Palfner (2003)).

33   See Walter (1990) and the transcripts of Duncker's wartime lectures held in the Bremen library (transcribed in part by M. Birkmann).

34   From Walter (1990) and H. Walter, pers. comm. and M. Birkmann, pers. comm. Walter never met Duncker personally; he talked to Dr H. Focke, who had been taught by Duncker and who was Duncker's successor at the Übersee-Museum in Bremen, but never spoke to Duncker, and Walter did not see the transcript of Duncker's Allies interview.

35   Coats and Phelps (1985); Collar et al. (1992). There is also a red siskin recovery programme: see http://www.afa.birds.org/redsiskin.htm.

10. HONEST RED?

1   This is from the Allies interview; the copy of letter to Jean Linsdale in the Übersee-Museum, Bremen. Exchanges of scientific journals were common but this was a somewhat asymmetric arrangement; a personal copy of *Condor* – one of the top three North American scientific bird journals (established in 1893) – in exchange for Duncker's quirky *Vögel ferner Länder* for Linsdale's university library at Berkeley.

2   See Wagner (1957) and Stresemann (1962). Surprisingly, I could find no photographs or any other account of this celebration.

3   Transcript of Duncker's lecture in Übersee-Museum, Bremen.

4   Professor Gerd von Wahlert, pers. comm.

5   Letters from Steinbacher to Duncker and from Julius Henniger to Duncker in Ubersee-Museum, Bremen. Henniger (1962).

6   Newspaper clippings in Duncker's file in the Übersee-Museum, Bremen.

7   Gill (1955).

8   See Koudis (2000).

9   I have changed some of the names here to protect the identity of the protagonists. For further details see Neslen (1978, p. 103), Lynch (1971, p. 106), Walker and Avon (1987).

10   Latscha (1990).

11   Rouche (1991). The problem was that each canthaxanthin tablet contained five times the safe daily dose and the tan advertisers recommended taking several tablets a day. They were sold under a variety of trade names and advertised widely, usually by showing two bikini-clad women side by side, one of them lily-white, the other temptingly tanned. To be fair, some companies did issue a warning about the use of canthaxanthin, although hardly of the right kind. One of them, for example, said, 'Both men and women have reported cases of too much attention from the opposite sex after taking Darker Tan™.' The poor woman who died was just twenty years old and when the doctor first saw her she was 'sitting propped up in bed, and what I could see of her – of her skin – was an awful shade of orange. So that was her wonderful tanning-parlour tan! It was inhuman. Frankly, she

was a sight. It was so pathetic. She had pale-red hair of the kind that goes with the fairest complexion, almost translucent complexion, and big pale-blue eyes, framed by that awful shade of orange.' Note that canthaxanthin is probably not toxic to birds and that the young woman who died after taking tanning tablets had numerous complications, including the fact that she was a Jehovah's Witness and refused a blood transfusion.

12   Scott (1934): *The Art of Faking Exhibition Poultry.*
13   Hill (2002).
14   Mayr (1984).
15   Compare and contrast Rose et al.'s (1984) *Not in Our Genes* with Alcock's (2001) *The Triumph of Sociobiology.*
16   See Segerstråle (2000); Deichmann (1996); Galton (2002).

POSTSCRIPT

1   The book by Walker and Avon (1987) is the best English guide to the current range of coloured canaries.
2   Stradi (1998); Stradi et al. (2001).

Bibliography

Aitinger, J. C., *Kurtzer und Einfältiger Bericht vom dem Vogelstellen*, Cassel: Johann Schützen, 1626.

Albin, E. *A Natural History of Birds* 3 vols. London. 1731–38.

Albin, E., *A History of English Song-birds*. London: Bettersworth & Co., 1737.

Alcock, J., *The Triumph of Sociobiology*, Oxford: Oxford University Press, 2001.

Aldrovandi, *Ornithologiae* hoc de avibis historiae libri xii. Bologna. Vol. i., 1599–1609.

Allen, G. E., *Thomas Hunt Morgan: The Man and his Science*, Princeton: Princeton University Press, 1978.

Andrewes, T., *The British Aviary*, London: William Cole, 1825.

—— *The Bird-Keeper's Guide and Companion*, London: Dean & Sons, 1830.

Anon., *The Bird Fancyer's Delight: Or Choice Observations and Directions concerning the feeding, breeding and teaching all sorts of Singing Birds*, London: Thomas Ward, 1714.

Anon., *The Bird-Fancier's Recreation: Being Curious Remarks on the Nature of Song-Birds with choice instructions concerning The taking, feeding, breeding and teaching them, and to know the Cock from the Hen*, London: privately published, 1728.

Anon., *A New Way of Breeding Canary Birds by a person who has bred canary birds for many years*, London: J. Hughes, 1742.

Anon., *Unterricht von den verschiedenen Arten der Kanarievögel und der Nachtigallen, wie diese beyderley Vögel aufzuziehen und mit Nützen so zu paaren seien, dass man schöne Zunge von ihnen haben kann*, Frankfurt: 1772.

Anon., 'Annual shows of the canary fancy', *Illustrated London News*, 12 December 1846, pp. 373–4.

Anon., *Canaries – Their Varieties and Points*, London: Dean & Son, 1873.

Anon., 'In the courts', *Bird Notes and News*, 1, 1905, pp. 47–8.

Anon., 'Herrn Generalkonsul C. H. Cremer – Bremen zum 16, August 1928', *Vögel ferner Länder*, 2, 1928, pp. 136–40.

Anon., Annual General Meeting, *Minutes of the Budgerigar Society (UK)*, 1929.

Anon., *D'après Nature: Chefs-d'oeuvre de la peinture naturaliste en Alsace de 1450–1800*, Strassbourg: Creamuse, 1994.

Armitage, L., *Science in Color Breeding*, Chicago: American Canary Magazine, 1947.

Armstrong, E. A., *The Folklore of Birds*, London: Collins, 1958.

Arnaiz-Villena, A., Alvarez-Tejado, M., Ruiz-del-Valle, V., Garcia-de-la-Tour, C., Varela, P., Recio, M. J., Ferre, S., Martinez-Laso, J., 'Phylogeny and rapid northern and southern hemisphere speciation of goldfinches during the Miocene and Pliocene epochs', *Cell. Mol. Life Sci.*, 54, 1998, pp. 1031–41.

Arnault de Nobleville, L. D., *Aedologie, ou Traite du Rossignol Franc, ou Chanteur*, Paris: Debure L'aine, 1751.

Arnold, M. L. *Natural Hybridisation and Evolution*. Oxford: Oxford University Press, 1997.

Astley, H. D., 'The hooded siskin and the wild canary', *Avic. Mag.*, 8, 1902, pp. 123–4.

Baines, A., *The Oxford Companion to Musical Instruments*, Oxford: Oxford University Press, 1992.

Ball, P., *Bright Earth: The Invention of Colour*, London: Viking, 2001.

Banks, R. C., 'Wildlife Importation into the United States, 1900–1972', US Dept. of Interior, Fish and Wildlife Service, Special Scientific Report, No. 200, Washington, DC, 1976.

Barbagli, F., Violani, C., 'Canaries in Tuscany', *Boll. Mus. reg. Sci. nat. Torino*, 15, 1997, pp. 25–33.

Barnesby, G. J., *The Canary: Its Management, Habits, Breeding and Training*, London: G. Routledge & Sons, 1877.

Barrett, P. H., Gautrey, P. J., Herbert, S., Kohn, D., Smith, S., *Charles*

Darwin's Notebooks, 1836–1844, London: British Museum/Cambridge University Press, 1987.

Barrington, D. H., 'Experiments and Observations on the singing of Birds', *Phil. Trans. Roy. Soc. Lond.*, lxiii, 1773, pp. 249–91.

Bateson, W., *Mendel's Principles of Heredity*, Cambridge: Cambridge University Press, 1902.

Bechstein, J. M., *Natural History of Cage Birds*, London: Groombridge, 1795.

Belon, P., *L'Histoire de la Nature des Oyseaux*. Paris, 1555.

Bemrose, E., Orme, 'How to obtain high-coloured canaries', *J. Horticulture & Cottage Gardener*, December 1873, p. 477.

Bennett, C. B., *A Study on the Nature of the Orange Canary*, Berkeley: privately published, 1947.

Bergmann, H. H., *Der Buchfink: Neues über einen bekannten Sänger*, Wiesbaden: Aula, 1993.

Berthold, P., *Bird Migration: A General Survey*, Oxford: Oxford University Press, 1993.

Bewick, T., *A Memoir of Thomas Bewick: Written by Himself*, Oxford: Oxford University Press, 1862.

BirdLife International & Council, E. B. C., *European Bird Populations: Estimates and Trends*, Cambridge: BirdLife International, 2000.

Birkhead, T. R., *Promiscuity: An Evolutionary History of Sperm Competition and Sexual Conflict*, London: Faber & Faber, 2000.

Birkhead, T. R., *The Wisdom of Birds*, London: Bloomsbury, 2010

Birkhead, T. R. & Charmantier, I., 'Nicolas Venette's Traite du rossignol (1697) and the discovery of migratory restlessness', *Archives of Natural History*, 40, 2013, pp. 125–138.

Birkhead, T. R., Giusti, F., Immler, S. & Jamieson, B. G. M., 'Ultrastructure of the unusual spermatozoon of the Eurasian bullfinch (*Pyrrhula pyrrhula*)', *Acta Zoologica*, 88, pp. 119-128.

Birkhead, T. R., Hall, J., Schut, E. & Hemmings, N., 'Unhatched eggs: methods for discriminating between infertility and early embryo mortality', *Ibis*, 2008b., pp. 150, 508–517.

Birkhead, T. R., Hemmings, N., Schut, E. & Fitzpatrick, S., 'Why bullies miss the bullseye', *Cage & Aviary Birds*, June 5 2008, p. 15.

Birkhead, T. R. & van Balen, S., 'Unidirectional hybridisation in birds: an historical review of bullfinch (*Pyrrhula pyrrhula*) hybrids', *Archives of Natural History*, 34, 2007, pp. 20–29.

Birkhead, T. R., Schulze-Hagen, K., Kinzelbach, R., 'Domestication of the canary *Serinus canaria* – the change from green to yellow', *Archives of Natural History*, 2004. pp. 50–56.

Birkhead, T. R., Schulze-Hagen, K., Palfner, G., 'The colour of birds: Hans Duncker, pioneer bird geneticist', *J. Ornithol.* 144, 2003, pp. 243–70.

Blagrave, J., *Epitome of the Art of Husbandry*, 1675.

Blakston, W. A., 'At last!', *J. Horticulture & Cottage Gardener*, 11 December 1873, pp. 476–7.

Blakston, W. A., Swaysland, W., Wiener, A. F., *The Book of Canaries and Cage Birds*, London: Cassell & Co., 1870.

Bowler, P. J., *Evolution: the History of an Idea*, Berkeley: University of California Press, 1983.

Brent, B. P., *The Canary, British Finches, and some other Birds*, London: *Journal of Horticulture & Cottage Gardener*, 1864.

Brockman, H., Völker, O., 'Der Gelbe Federfarbstoff des Kanarienvogels (Serinus canaria canaria (L.)) und das Vorkommen von Carotinoiden bei Vögeln', *Hoppe-Seyler's Zeitschrift für physiologische Chemie*, 224, 1934, pp. 193–215.

Broeckhoven, M., *De Vinkensport in Vlaanderen*, Gent: Koninklijke Bond der Oostvlaamse Volkskundigen, 1969.

Browne, J., *Charles Darwin: Voyaging*, London: Jonathan Cape, 1995.

Bub, H., *Bird Trapping and Bird Banding*, Ithaca: Cornell University Press, 1991.

Buffon, G. L., *The Natural History of Birds from the French* of Count Buffon in 9 volumes. London: Strahan Cadell & Murray, 1793.

Burkhardt, R. W. J., 'Closing the door on Lord Morton's Mare: The rise and fall of telegony', *Studies in the History of Biology*, 3, 1979, pp. 1–21.

Burleigh, M., *The Third Reich*, London: Macmillan, 2000.

Calegari, S., Radici, F., Mora, T. d. V., *I Roccoli Della Bergamasca*, Realizzazione editorial, 1985.

Carluccio, A. and Carluccio, P., *Carluccio's Complete Italian Food*, London: Quadrille, 1997.

Coats, S., Phelps, W. H. J., 'The Venezuelan red siskin: Case history of an endangered species', *Ornithological Monographs*, 36, 1985, pp. 977–85.

Collar, N. J., Gonzaga, L. P., Krabbe, N., Madrono Nieto, A., Naranjo, L. G., Parker, T. A. I., Wege, D. C., *Threatened Birds of the Americas: The ICBP/IUCN Red Data Book*, Cambridge, UK: International Council for Bird Preservation, 1992.

Cookson, G., 'Propagation by hybrids', *Annals of Natural History*, 5, 1840, pp. 424–5.

Copeland, P., Boswall, J., Petts, L., *Birdsongs on Old Records*, London: British Library National Sound Archive Wildlife Section, 1988.

Cox, N., *The Gentleman's Recreation*, London: E. Flesher, for Maurice Atkins at the Half-moon in St Paul's Church-yard, and Nicolas Cox over against Furnivals-Inn-Gate in Holborne, 1677.

Crew, A. F. E., Lamy, R., 'Genetics and the budgerigar', *Budgerigar Bulletin*, 32, 1934, pp. 161–6.

—— *The Genetics of the Budgerigar*, London: Watmough, 1935.

Curtis, A. T., Speakman, F. J., *A Poacher's Tale*, London: Bell & Sons, 1960.

Dams, A., Bemerkenswerte Neuzüchtungen [Remarkable new breeds], *Gefiederte Welt*, 54, 1925, pp. 424–6.

—— Farbenkanarien [Coloured Canaries], *Gefiederte Welt*, 55, 1926, pp. 535–8.

Damsté, P. H., 'Experimental modification of the sexual cycle of the green-finch', *J. Exp. Biol.*, 24, 1947, pp. 20–35.

Darwin, C., *On the Origin of Species*, London: John Murray, 1859.

—— *The Variation of Animals and Plants under Domestication*, London: John Murray, 1868.

—— *The Descent of Man, and Selection in Relation to Sex*, London: John Murray, 1871.

Davenport, C. B., 'Inheritance in Canaries', *Carnegie Institute of Washington*, 95, 1908, pp. 5–26.

—— 'Dr. Galloway's "Canary Breeding" ', *Biometrika*, 7, 1910, pp. 398–400.

Dawkins, R., *The Selfish Gene*, Oxford: Oxford University Press, 1976, 1989.

—— *The Extended Phenotype*, Oxford: Oxford University Press, 1982.

Dean, W., 'Is there yellow in hooded siskins?', *Cage & Aviary Birds*, 11 July 1963, pp. 28–9.

Deelder, C. L., 'Gegevens over bloemendaalse vinkenbaan', *Ardea*, 39, 1951, pp. 321–41.

Deichmann, U., *Biologists Under Hitler*, Cambridge, Mass.: Harvard University Press, 1996.

Diamond, J., *Guns, Germs and Steel*, London: Jonathan Cape, 1997.

Dixon, E. S., *The Dovecote and the Aviary*, London: Orr & Co., 1851.

Duncker, H., 'Über die Homologie von Cirrus und Elytron bei den Aphroditiden (Ein Beitrag zur Morphologie der Aphroditiden)', *Zeitschr. f. Wiss. Zool.*, 81, 1905a, pp. 191–276.

——— *Wanderzug der Vögel*, Jena: Petsche-Labarre-Stiftung, 1905b.

——— 'Nachtigall-Kanarienhähne', *Kosmos*, 5, 1922a, pp. 129–30.

——— 'Die Reich'sche Gesangeskreuzung (Nachtigall u. Kanarienvogel) eine "erworbene" Eigenschaft?', *J. Ornithol.*, 70, 1922b, pp. 423–30.

——— 'Die vogelhäuser von Herrn Generalkonsul C. H. Cremer, Bremen', *Vögel ferner Länder*, 1, 1927a, pp. 166–74.

——— 'Bastarde von Kapuzenzeisig und weissen Kanarievögel', *Vögel ferner Länder*, 1, 1927b, pp. 67–74.

——— 'Genetik der Kanarienvögeln', *Bibl. Genet.*, 4,1928, pp. 40–140.

——— 'Genetik der Wellenstittiche', *Der Züchter*, 2, 1929a.

——— *Kurzgefasste Vererbungslehre für Kleinvogel-Züchter*, Leipzig: Dr F. Poppe, 1929b.

——— 'Farbenvererbung bei Buntvögeln', *Vögel ferner Länder*, 3, 1929c, pp. 90–109.

——— 'Röntgenstrahlen und Keimschädigung', *Strahlentherapie*, 37, 1930, pp. 142–63.

——— 'Generalkonsul Carl Hubert Cremer', *Gefiederte Welt*, 15, 1938a, pp. 169–71.

——— 'Farbenzucht. III. Die Karthothekkarte', *Kanaria 1938* (1), 1938b, pp. 4–5.

Duncker, H., Lange, F., *Neue Ziele und Wege des Biologieunterrichts. Vier Beiträge*, Frankfurt am Main: Diesterweg, 1934.

Durham, F. M., 'Sex-linkage and other genetical phenomena in canaries', *Journ. Genet.*, XVI, 1926, pp. 19–33.

Durham, F. M., Marryat, D. C. E., 'Note of the inheritance of sex in canaries', *Rep. Evol. Comm. Roy. Soc.*, IV, 1908, pp. 57–60.

Ellegren, H., Lindgren, G., Primmer, C. R., Møller, A. P., 'Fitness loss and germline mutations in barn swallows breeding in Chernobyl', *Nature* (London), 389, 1997, pp. 593–6.

Evans, A. H., *Turner on Birds*, Cambridge: Cambridge University Press, 1903.

Evans, H., 'The Dawn of an Era', *Lizard News*, parts 1 and 2, 1996, no page numbers.

Evans, W. E., *The Songs of the Birds*, London: Francis & John Rivington, 1845.

Fenech, N., *Fatal Flight: The Maltese Obsession with Killing Birds*, London: Quiller Press, 1992.

—— 'Bird Shooting and Trapping in the Maltese Islands: Some Socio-Economic, Cultural, Political, Demographic and Environmental Aspects', unpublished Ph.D. thesis, University of Durham, UK, 1997.

Fisher, J., *Zoos of the World*, London: Aldus, 1966.

Fortin, F., *Les Ruses Innocentes*, 1660.

Fumihito, A., Miyake, T., Takada, M., Shingu, R., Endo, T., Gojobori, T., Kondo, N., Ohno, S., 'Monophyletic origin and unique dispersal patterns of domestic fowls', *Proc. Nat. Acad. Sci.*, 93, 1996, pp. 6792–5.

Gaddis, T. E., *Birdman of Alcatraz*, New York: Aeonian Press, 1955.

Galloway, R., 'Canary Breeding. A partial analysis of records from 1891–1909', *Biometrika*, 7, 1909, pp. 1–42.

—— 'Canary Breeding. A Rejoinder to C. B. Davenport', *Biometrika*, 7, 1910, pp. 401–3.

—— 'History of the Canary' in Robson, J., Lewer, S. H. (eds), *Canaries, Hybrids and British Birds in Cage and Aviary*, London: Waverley Books, 1911.

Galton, D., *Eugenics*. London: Abacus 2002.

Galton, F., 'The first steps towards the domestication of animals', *Transactions Ethnological Society London*, 3, 1865, pp. 122–38.

Galton, F., *Hereditary Genius*, London: Macmillan, 1869.

Garthus, M., *The Way We Lived in Germany during World War II*, Perth: Arale Books, 1977.

Gasser, C., 'Die Imster Vogelhändler', *Der Schlern, Monatszeitschr. f. Südtiroler Landeskunde*, 75, 2001, pp. 992–1008.

Gattiker, R., Gattiker, L., *Die Vögel im Volksglauben*, Wiesbaden: Aula, 1989.

Gaunt, A. S., Wells, M. K., 'Models of syringeal mechanisms', *American Zoologist*, 13, 1973, pp. 1227–47.

Gessner, C., *Historiae Animalium* libre III, Zurich, 1555.

Gill, A. K., *New-coloured Canaries*, London: Cage Birds, 1955.

Gillham, N. W., *A Life of Sir Francis Galton*, Oxford: Oxford University Press, 2001.

Gleich, O., Klump, G M & Dooling, R J. 'Peripheral basis for the auditory deficit in Belgian Waterslager canaries (*Serinus canarius*)', Hearing Research, 82, 1995, pp. 100–108.

Godman, S., 'The Bird Fancyer's Delight', *Ibis*, 97, 1955, pp. 240–6.

Goodwin, D., *Instructions to Young Ornithologists*, London: Museum Press Ltd, 1965.

Grant, P. R., Grant, B. R., 'Hybridisation of bird species', *Science*, 256, pp. 193–7.

Gray, A. P., *Bird Hybrids: a check-list with bibliography*, Commonwealth Agricultural Bureaux, Farnham Royal, UK, 1958.

Gray, P. M., Krause, B., Aterna, J., Payne, R., Krumhansl, C., Baptista, L., 'The music of nature and the nature of music', *Science*, 291, 2001, pp. 52–4.

Grenze, v. d. H., 'Die Nachtingall-Edelkanarien–Karl Reich–Bremen über ein Lebenswerk' [The nightingale-Edelkanarien–Karl Reich–Bremen on his life and achievements], *Kanaria*, week 30, 1938, pp. 350–2.

Grinnell, J., 'The linnet of the Hawaiian Islands: a problem in speciation', *University of California Publications in Zoology*, 7, 1911, pp. 179–95.

Güttinger, H. R., 'Consequences of domestication on the song structures in the canary', *Behaviour*, 94, 1985, pp. 255–78.

Güttinger, H. R., Turner, T., Dobmeyer, S., Nicolai, J., 'Melodiewahrnehmung und Wiedergabe beim Gimpel: Untersuchungen an liederpfeifenden und Kanariengesang imitierenden Gimpeln (*Pyrrhula pyrrhula*)', *J. Ornithol.*, 143, 2002, pp. 303–18.

Hackethal, S., 'Betrachtungen zur Tierdarstellung in der Renaissance anhand der Aquarelle von Lazarus Röting (1549–1614)', *Gesch. Naturw. Techn. Med.*, 27, 1990, pp. 49–64.

Haeckel, E., *The Riddle of the Universe*, London: Watts & Co., 1900.

—— *The Wonders of Life*, London: Watts & Co., 1905.

Haffer, J., 'Erwin Stresemann (1889–1972) – Life and work of a pioneer in scientific ornithology: a survey', *Acta Historica Leopoldina*, 34, 2000, pp. 399–427.

—— 'Ornithological research traditions in central Europe during the 19th and 20th centuries', *J. Ornithol.*, 142, 2001, pp. 27–93.

Haldane, J. B. S., 'Sex ratio and unisexual sterility in hybrid animals', *Journal of Genetics*, 12, 1922, pp. 101–9.

Hamersley, J., *The Bird Fancyer's Delight*, London, 1717.

Harley, R. D., *Artists' Pigments c. 1600–1835*, London: Camelot Press, 1982.

Harting, J. E., *The Ornithology of Shakespeare*, London: Van Voorst, 1871.

Henig, R. M., *A Monk and Two Peas: The Story of Gregor Mendel and the Discovery of Genetics*, London: Weidenfeld & Nicolson, 2000.

Henniger, J., *Farbenkanarien*, Leipzig & Berlin: Maximiliansau, 1962.

Heron, D., 'Inheritance in Canaries: A study in Mendelism', *Biometrika*, 7, 1910, pp. 403–10.

Herrick, E. H., Harris, J. O., 'Singing female canaries', *Science*, 125, 1957, pp. 1229–30.

Hervieux de Chanteloup, J.-C. C., *Nouveau Trait des Serins de Canarie* (English translation 1719; *A New Treatise of Canary Birds*), Paris: Claude Prodhomme (English translation) Bernard Lintot, 1708.

Hill, G., *A Red Bird in a Brown Bag*, Oxford: Oxford University Press, 2002.

Hinkelmann, C., 'Alle guten Dinge sind drei – die Herausgeber der Gefiederten Welt', *Gefiederte Welt*, May 2001, pp. 148–54.

Hohberg, W. H. von., *Waidmannschaft Durchs Gantze Jahr (The Hunter's Year)*, 1703.

Hölzinger, J., *Die Vögel Baden-Württembergs. Gefährdung und Schutz*, Stuttgart: Ulmer, 1987.

Hoos, D., 'Vinkenbaan: Hoe het er toeging en wat er mee in verband stond', *Ardea*, 26, 1937, pp. 173–202.

Hope, T., *The Fancier's Necessary Companion and Sure Guide. In two parts*, London: Thomas Hope, 1762.

Hopkinson, E., 'Hooded siskin mules', *Avicultural Magazine*, 9, 1920, p. 150.

Houlton, C., *Cage-bird Hybrids*, London: Cage-Birds, c.1920.

Huxley, J., *The Modern Synthesis*, London: Allen & Unwin, 1942.

Jackson, C. E., *Bird Etchings: The Illustrators and their books 1655–1855*, Ithaca: Cornell University Press, 1985.

Jongh, E. d., 'Erotica in vogelperspectief: De dubbelzinnigheid van een reeks 17-eeuwse genrevoorstellingen', *Simiolus*, 3, 1968–9, pp. 22–74.

Keeler, C. A., 'Evolution of Colors in North American Land Birds', *Occasional Papers California Academy of Sciences*, III, 1893a.

——— 'The Evolution of the Colors of North American Birds – A Reply to Criticism', *The Auk*, 10, 1893b, pp. 373–7.

Keynes, R., *Annie's Box*, London: Fourth Estate, 2001.

King-Hele, D., *Erasmus Darwin*, London: De la Mare, 1999.

Kinzelbach, R. K., Holzinger, J., *Marcus zum Lamm (1544–1606): Die Vogelbücher aus dem Thesaurus Picturarum*, Stuttgart: Eugen Ulmer, 2001.

Knolle, F., *Mensch und Vogel im Harz*, Clausthal-Zellerfeld: Piepersche Druckerei und Verlagsanstalt, 1980.

Koudis, S., *Canary Colour Breeders Millennium Limited Edition 1947–1999*, Canary Colour Breeders Association, 2000.

Kragenow, P., *De Buchfink*, Wittenberg: Lutherstadt, 1981.

Kroodsma, D. E., 'Reproductive development in a female songbird: Differential stimulation by quality of male song', *Science*, 192, 1976, pp. 574–5.

Kroll, M., *Letters from Liselotte*, London: Allison & Busby, 1998.

Kuthy, R., 'Vogelfang in Österreich – Unter besonderer Berücksichtigung des Salzkammergute. Der Weg von Kultureller Universalität ins "out" ' (Bird-catching in Austria, with special attention to the Salzkammergut area: the way from universal culture to being 'out of fashion'), PhD: thesis, University of Vienna, 1993.

LaBruyère, *Caractères*, 1688.

Lander, P., Partridge, B., *Popular British Birds in Aviculture*, Waterlooville: Kingdom Books, 1998.

Latscha, T., *Carotenoids: Their Nature and Significance in Animal Feeds*, Basel: Hoffman-La Roche, 1990.

Legendre, M., 'Histoire de l'origine des canaries', *L'Oiseau et la revue Française d'Ornithologie*, 1955a, pp. 185–98.

——— *Le Serin de Canaries: Historique, élevages, races, hybrides, nourriture colorante*, Paris: Editions Boubée, 1955b.

Leitner, S., Voigt, C., Gahr, M., 'Seasonal changes in the song pattern of

the non-domesticated island canary (*Serinus canaria*), a field study',
Behaviour, 138, 2001, pp. 885–904.

Lipset, D., *Gregory Bateson: The Legacy of a Scientist*, Englewood Cliffs,
New Jersey: Prentice-Hall, 1980.

Loisel, G., *Histoire des Ménageries*, Paris: Octave Doin et Fils & Henri
Laurens, 1912.

Lynch, G., *Canaries in Colour*, Poole: Blandford, 1971.

MacPherson, H. A., 'The Goldfinch', *Midland Naturalist*, IV, 1880, pp. 225–33.

—— *A History of Fowling*, Edinburgh: D. Douglas, 1897.

Major, R. H., *The Canarian, or Book of the Conquest and Conversion of the
Canarians*, London: Hakluyt Society, 1872.

Manzini, C., *Ammaestramenti per allevare, pascere, et curare gli uccelli*,
Brescia: Pietro Maria Marchetti, 1575.

Markham, G., *Hungers Prevention: or The Whole Art of Fowling by Water
and Land*, London: Francis Grove, 1621.

Matthey, I., *Vincken moeten vincken locken. Vijf eeuwen vangst van zangvo-
gels en kwartels in Holland*, Hilversum. Historische, Vereniging Holland,
2002.

Mayhew, H., *London Labour and the London Poor*, London: Griffin, Bohn,
1861.

Mayr, E., 'Germany' in *The Evolutionary Synthesis* (Mayr, E., Provine, W.
B. eds), Cambridge, Mass.: Harvard University Press, 1980, pp. 279–84.

—— *The Growth of Biological Thought: Diversity, Evolution, and
Inheritance*, Cambridge, Mass.: Belknap Press, 1982.

—— Introduction in *Perspectives in Ornithology* (Brush, A. H., Clark, G.
A. J. eds), Cambridge: Cambridge University Press, 1984, pp. 1–21.

Medawar, P. B., Medawar, J. S., *Aristotle to Zoos*. London: Weidenfeld &
Nicolson, 1984.

Meyer, H., Duncker, H., *Von der Verhütung unwerten Lebens: Ein Zyklus
von 5 Vortragen*, Bremen: G. A. v. Halem, 1933.

Miller, G., *The Mating Mind*, London: Heinemann, 2000.

Milon, P., 'Notes Bibliographiques sur le Nouveau Trait des Serins de
Canarie D'Hervieux de Chanteloup', *Oiseau*, 26, 1956, pp. 204–18.

Montagu, G., *Ornithological Dictionary*, London, 1831.

Müller-Hill, B., *Murderous Science*, Oxford: Oxford University Press, 1988.

Mundinger, P. C., 'Behaviour-genetic analysis of canary song: interstrain differences in sensory learning, and epigenetic rules', *Animal Behaviour*, 50, 1995, pp. 1491–511.

Munro, G. C., *Birds of Hawaii*, Rutland: Charles E. Tuttle Co. Inc., 1944.

Napoli, J. F., 'Denazification from an American's viewpoint', *Annals of the American Academy of Political and Social Science*, 1949, pp. 115–23.

Neslen, J. M., *All About Canaries*, London: Barrie & Jenkins, 1978.

Neslen, J. M., *Introduction to the New Colour Canary*, Liss, Hants.: Nimrod, 1989.

Newton, A., 'Canary-Bird' in *Dictionary of Birds*, London: Black, 1896, pp. 70–2.

Newton, I., *Finches*, London: Collins, 1972.

Nicolai, J., 'Zur Biologie und Ethologie des Gimpels (*Pyrrhula pyrrhula* L)', *Z. Tierpsychologie*, 13, 1956, pp. 93–132.

—— 'Folksong singing bullfinches' in *American Federation of Aviculture*, Miami, USA, 1993.

Nyhart, L. K., *Biology Takes Form: Animal Morphology and the German Universities 1800–1900*, Chicago: University of Chicago Press, 1995.

Ohno, S., *Sex Chromosomes and Sex Linked Genes*, Berlin: Springer-Verlag, 1967.

Olina, G. P., *Uccelliera*, Rome: Andrea Fei, 1622.

Orr, H. A., 'Haldane's Rule', *Annual Review of Ecology & Systematics*, 28, 1997, pp. 195–218.

Osman, H. J., *The Orange Canary*, Oakland, California: privately published, 1948.

Page, W. J., 'Death of Mr A. K. Gill robs fancy of a great pioneer', *Cage & Aviary Birds*, 1957.

Page, W. T., *Species which have reared young and Hybrids which have been bred in captivity in Great Britain*, Ashbourne: The Avian Press, 1914.

Parsons, J. J., 'The origin and dispersal of the domesticated canary', *J. Cultural Geography*, 7, 1987, pp. 19–34.

—— 'A detailed history of the domesticated canary', *American Cage-Birds Magazine*, March 1989, pp. 17–30.

Pauly, P. J., *Biologists and the Promise of American Life*, Princeton: Princeton University Press, 2000.

Pernau, V., *Angenehmer Zeit-Vertreid* (1716), cited in Stresemann (1947).

Potts, M., Short, R. V., *Ever Since Adam and Eve*, Cambridge: Cambridge University Press, 1999.

Price, T., 'Domesticated birds as a model for the genetics of speciation by sexual selection', *Genetica*, 116, 2002, pp. 311–27.

Price, T. D., Bouvier, M. M., 'The evolution of F1 postzygotic incompatibilities in birds', *Evolution*, 56, 2002, pp. 2083–9.

Promptova, A. N., 'Hybridisation of Fringillidae', *J. Biol. Exp.* (in Russian; English abstract: *Biological Abstracts*, 1930), 4, 1928, p. 640.

Provine, W. B., *The Origins of Theoretical Population Genetics*, Chicago: University of Chicago Press, 1971.

Raven, C. E., *English Naturalists from Neckam to Ray*, Cambridge: Cambridge University Press, 1947.

Ray, J., *The Ornithology of Francis Willughby* (1676), London: John Martyn, 1678.

Rettich, A., 'A chaffinch singing match', *Avicultural Magazine*, 2,1896, pp. 114–17.

Ridley, M., *Nature via Nurture*, London: Fourth Estate, 2003.

Ringleben, H., 'Lebensskizzen von Ornithologen im Lande Bremen', *Abh. Naturw. Verein Bremen*, 43, 1955, pp. 5–28.

Ritvo, H., *The Animal Estate*, Harvard: Harvard University Press, 1987.

Robbins, L. E., *Elephant Slaves and Pampered Parrots: Exotic Animals in Eighteenth-Century Paris*, Baltimore: John Hopkins University Press, 2002.

Roberts, S., *Bird-Keeping and Birdcages: A History*, Newton Abbott: David & Charles, 1972.

Robson, J., Lewer, S. H., *Canaries, Hybrids and British Birds in Cage and Aviary*, London: Waverley Books, 1911.

Robson, J., *Canary Management Throughout the Year*, London: The Feathered World, 1918.

Rose, S., Lewontin, S., Kamin, L., *Not in Our Genes*, London: Penguin, 1984.

Samstag, T., *For Love of Birds: The Story of the Royal Society for the Protection of Birds, 1889–1988*, Sandy, Beds: RSPB, 1988.

Rouche, B., 'Annals of medicine: A good, safe tan', *New Yorker*, 11 March 1991, pp. 69–74.

Santens, F., *Hinke de Vinke: 400 jaar Vinkensport in Vlaanderen: 60 jaar Avibo*, Vichte: privately published, Printer Vanoverbeke, 1995.

Schiavone, M., 'Due rare opere di letteratura venatoria', *Rivista Italiana di Ornitologia*, 48, 1978, pp. 239–41.

Schlenker, R., 'Johann Ferdinand Adam Freiherr von Pernau (1660–1731): Beiträge zu einer Bibliographie seiner vogelkundlichen Schriften', *Jahrbuch der Coburger Landesstiftung*, 27, 1982, pp. 225–38.

—— 'Johann Matthäus Bechstein (1757–1822): ein Beitrag zu einer Bibliographie seiner Schriften', *Anz. Ver. Thüring. Ornithol.*, 2, 1994, pp. 125–33.

Schwerdt, R., *Hunting, Hawking, Shooting illustrated in a Catalogue of Books, Manuscripts, Prints and Drawings*, London: Waterlow, 1928.

Scott, G. R., *The Art of Faking Exhibition Poultry*, London: Werner & Laurie, 1934.

Schulze-Hagen, K., Geus, A., *Joseph Wolf (1820–1899): Animal Painter*, Marburg: Basilisken Presse, 2000.

Secord, J. A., 'Nature's Fancy: Charles Darwin and the Breeding of Pigeons', *Isis*, 72, 1981, pp. 163–86.

—— 'Darwin and the Breeders: A Social History' in *The Darwinian Heritage* (Kohn, D. ed.), Princeton: Princeton University Press, 1985, pp. 519–42.

Segerstråle, U., *Defenders of the Truth*, Oxford: Oxford University Press, 2000.

Sheldon, B. J., Burke, T., 'Copulation behaviour and paternity in the chaffinch', *Behavioural Ecology & Sociobiology*, 34, 1994, pp. 149–56.

Solinas, F., *l'Uccelliera: Un libro di arte e di scienza nella roma dei primi lincei*, Florence: Leo S. Olschki Editore, 2000.

Sossinka, R., 'Domestication in birds' in *Avian Biology*, (eds. King, A. S., McLelland, J.) Academic Press, 1982, pp. 373–403.

Specter, M., 'Rethinking the brain', *New Yorker*, 23 July 2001, pp. 42–53.

Speicher, K., 'Farbem Kanarien: Benennung der Spielarten in der Farbkanarienzucht', *AZ Journal*, 1962, pp. 109–10.

Speicher, K., *Kammersänger im Federkleid: Kanarievögel*, Stuttgart: Franckesche Verlagshandlung, 1976.

Spiller, J., *Klee*, London: Blandford, 1962.

Stein, G. J., 'Biological science and the roots of Nazism', *American Scientist*, 76, 1988, pp. 50–8.

Stong, C. L., 'The Amateur Scientist: The strange story of Robert Stroud, who studied birds while in solitary confinement', *Scientific American*, 197, 1957, pp. 143–53.

Stradi, R., *The Colour of Flight*, Milan: Solei Gruppo Editoriale Informatico, 1998.

Stradi, R., Pini, E., Celentano, G., 'Carotenoids in bird plumage: the complement of red pigments in the plumage of wild and captive bullfinch (*Pyrrhula pyrrhula*)', *Comparative Biochemistry & Physiology*, 2001, pp. 529–535.

Stresemann, E., 'Die Vogelbilder des Nürnbergers Lazarus Röting' (1614), *Verhandl. der Ornithol. Ges. Bayern*, 15, 1923a, pp. 308–15.

— 'Zur Geschichte einiger Kanarievogel-Rassen', *Ornith. Monatsber.*, 31, 1923b, p. 103.

—— 'Baron von Pernau, pioneer student of bird behaviour', *The Auk*, 64, 1947, pp. 35–52.

Dr Hans Duncker. *J. Ornithol.* 103, 1962, p. 122.

—— *Ornithology from Aristotle to the Present*. Harvard: Harvard University Press, 1975.

Swainson, W., *Zoological Illustrations or Original Figures and Descriptions of New, rare, or Interesting Animals selected chiefly from the classes of ornithology, entomology and conchology*, vol. 1, London, 1820.

Syme, P., *A Treatise on British Songbirds*. Edinburgh: John Anderson, 1823.

Tappen, J., 'Kan er een einde komen aan de enorme vogelvangst in Italië?' *Overdruk uit De Zwerver*, August/September 1959, pp. 127–30.

Temmen, *Vom Dritten Reich zum Wirtschaftswunder: Bremen und Bremerhaven 1933–1955*, Bremen: Edition Temmen, 2001.

Thielcke, G., 'Neue Befunde bestätigen Baron Pernaus (1660–1731) Angaben über Lautäusserungen des Buchfinken (*Fringilla coelebs*)', *J. Ornithol.*, 129, 1988, pp. 55–70.

Thorpe, W. H., 'Comments on "The Bird Fancyer's Delight": together with notes on imitation in the sub-song of the chaffinch', *Ibis*, 97, 1955, pp. 247–51.

—— *Bird-Song*, Cambridge: Cambridge University Press, 1961.

Toth-Ubbens, M., 'Kijken naar een vogeltje' in *Miscellanea I. Q. van Regteren Altena*, Amsterdam, 1969, p. 155.

Vale, M. G. A., *Charles VII*, London: Methuen, 1974.

Vallet, E., Beme, I., Kreutzer, M., 'Two-note syllables in canary songs elicit high levels of sexual display', *Animal Behaviour*, 55, 1998, pp. 291–7.

Valli da Todi, A., *Il canto de gl'Augelli. Opera nova. Dove si dichiara la natura di sessanta sorte di Uccelli, che cantano per esperienza, e diligenza fatta piu volte. Con il modo di pigliarli con facilita, & allevarli, cibarli, domesticarli, ammaestrarli e guaririli delle infermita, che a detti possono succedre. Con le loro figure, o vinti sorte di caccie, cavate dal naturale da Antonio Tempesti*, Rome: N. Mutii, 1601.

Vins, T., *Das Wellensittichbuch*, Alfeld: M. & H. Schaper, 1993.

Wagner, H. O., 'Dr Hans Duncker', *Abh. Naturw. Verein Bremen*, 34, 1957, pp. 173–5.

Walker, G. B. R., Avon, D. *Coloured, Type and Song Canaries*, Poole: Blandford, 1987.

Walter, H., 'Die Rassenhygienische Fachgesellschaft (1931–1945) im Naturwissenschaftlichen Verein zu Bremen', *Abh. Naturw. Verein Bremen* 41, 1990, pp. 31–48.

Waterton, C., *Essays on Natural History*, London: Frederick Warne & Co., 1870.

Watmough, W., *The Cult of the Budgerigar*, London: Cage Birds, 1935.

Wells, C., 'The Early Flageolet', *Recorder Magazine*, December 1993, pp. 72–4.

Wetmore, A., 'Canaries and other cage-bird friends', *National Geographic Magazine*, 74, pp. 775–806.

Wiener, A. F., 'The hooded siskin', *Avic. Mag.* (Ser. 2), 1, 1903, pp. 115–16.

Wood, R. J., Orel, V., *Genetic Prehistory in Selective Breeding: A Prelude to Mendel*, Oxford: Oxford University Press, 2001.

Wright, R., *The Moral Animal*, London: Random House, 1994.

Yapp, W. B., *Birds in Medieval Manuscripts*, London: British Library, 1981.

—— 'Birds in captivity in the Middle Ages', *Archives of Natural History*, 10, 1982, pp. 479–500.

Zeraschi, H., *L'orgue de Barbarie*, Paris: Diffusion en France: éditions Payot, 1980.

Zeuner, F., *A History of Domesticated Animals*, London: Hutchinson, 1963.

Index

Piles, Roger de, 30
pipit, 18
Plato, 58
Polydore Vergil, 14
Prussia, 169

quagga, 144–5

rabbits, 92
racehorse breeding, 113
racial hygiene, 185, 186, 192–4
radiation, 112,130
Raleigh, Sir Walter, 32
Rastenburg, 169
rats, 90, 92
Ray, John, 86
Red Canary Movement, 168
red pepper, 128–9, 179, 200, 205, 208, 210
red siskin (*Carduelis cucullata*), 1, 135–6, 146,
    173, 175, 177, 215; colour change, 175,
    182–3; diet, 181; 'doppelbastards', 151, 154;
    endangered, 184, 196–7, 199; fashion for
    skins, 196; genes, 166, 215–6; mules, 136–8,
    142, 147–8, 151, 154–5, 163, 169; trade in,
    196–7
red wolf, 148
redpoll, 10, 96, 209, 215
reed bunting, 30
Reich, Karl: canary breeding, 32, 40, 42, 46,
    57, 69; recordings of birdsong, 39, 43–4;
    Duncker's meeting with, 39–40, 57, 93;
    shifts nightingale's singing season, 40–3;
    research on canary variegation, 72–4, 99,
    155, 215; and canary evolution, 101, 109;
    at Rosenau, 134; mule breeding, 151,
    154–5; dedicatee of Duncker's book, 170;
    politics unknown, 186; leaves Bremen,
    190
Riddle of the Universe, The (Haeckel), 162
Rinaldo (Handel), 63
robin, 26, 115
roccoli, 16, 17
rock dove (*Columba livia*), 101
Rockefeller University, 51
Romans, 83–4
rook, 54
Roosevelt, Franklin D., 190
Rosenau, 134, 153, 158
Rossitten, 12
Röting, Lazarus, 96–97, 97, 98, 137, 139, 190
Royal Society, 118, 145

Royal Society for the Protection of Birds
    (RSPB), 5
Russian comfrey, 204
Rustic Love (Jan Steen), 20

saffron siskin, 177
St Andreasberg, 35
San Francisco, 169, 175
scarlet tanager, 177
Schnorr, Ludwig, 47
Schroeckius, Lucas, 97, 98, 163
Schwan, Albrecht, 98
Schwartzburgbund, 11
Science in Color Breeding (Armitage), 174
sea mice, 11
Sebright, Sir John, 108
Second World War, 175, 212; outbreak of, 189
selective breeding, 155, 161
Selfish Gene, The (Dawkins), 211
sensitive period, 60
Segerstråle, U., quoted, 153
serin, 10, 80, 98, 139, 143, 147
serinette, 62–4
sex, 18–22
sex chromosomes, 150
sexual display, 56, 69
sexual imprinting, 67
sexual selection, 19–20, 21, 50, 51–2, 53, 176,
    208
sheep, 86, 87, 89, 91, 107, 113, 179; merino, 155,
    179
Shoreditch Bobby (chaffinch), 46
Siberia, 183
siffleur d'oiseaux, 79
Silbo Gomera, 33
siskin, 10; hybrid, 120, 143, 147; *see also* red
    siskin
skylark, 45, 63
social Darwinism, 109, 162, 195
sociobiology, 211–2
soma, 58
Sorel, Agnès, 18–19
Spain, 21, 33, 84
sparrow, 63
Spiller, Jürg, quoted, 153
Stalin, 190
starling, 63
Steen, Jan, 20
Stein, George, 162
Steinbacher, Joachim, 201–2
stereotypes, national, 28